T0184367

INTERNATIONAL CENTRE FOR MECHANICAL SCIENCES

COURSES AND LECTURES - No. 65

19. Juni 1975

DIETER BESDO
TECHNICAL UNIVERSITY OF BRUNSWICK

EXAMPLES TO EXTREMUM AND VARIATIONAL PRINCIPLES IN MECHANICS

SEMINAR NOTES ACCOMPANING THE VOLUME
No. 54 BY H. LIPPMANN

COURSE HELD AT THE DEPARTMENT
OF GENERAL MECHANICS
OCTOBER 1970

UDINE 1973

SPRINGER-VERLAG WIEN GMBH

This work is subject to copyright.

All rights are reserved,

whether the whole or part of the material is concerned

specifically those of translation, reprinting, re-use of illustrations,

broadcasting, reproduction by photocopying machine

or similar means, and storage in data banks.

Copyright 1972 by Springer-Verlag Wien
Originally published by Springer-Verlag Wien New York in 1972

ISBN 978-3-211-81230-3 ISBN 978-3-7091-2726-1 (eBook)
DOI 10.1007/978-3-7091-2726-1

P R E F A C E

The following examples to extremum and variational principles in mechanics were delivered in a seminar which accompanied a lecture course of Professor Horst LIPPMANN, Brunswick. Therefore, the examples cannot stand for themselves, their main function was to illustrate the results of the lecture course and to demonstrate several interesting peculiarities of the single solution methods.

The problems are normally chosen to be quite simple so that numerical computations are not necessary. Nevertheless, sometimes, the calculations will only be mentioned and not worked out here.

The sections of the seminar-course are not identical with those of the lecture course. Especially, there are no examples to more or less theoretical sections of the lectures. Because of the close connection to the lectures, no separate list of references is given. Also the denotation is mostly the same as in the lecture-notes.

I say many thanks to Professor Horst LIPPMANN for his help during the preparation-time and to the International Centre for Mechanical Sciences for the invitation to deliver this seminar.

Brunswick, October 31, 1970

Dieter Besdo

1. EXTREMA AND STATIONARITIES OF FUNCTIONS

1.1. Simple problems (cf. sect. 1.2 of the lecture-notes)

In this sub-section, several simple problems have to demonstrate definite peculiarities which may occur if we want to calculate extrema of functions.

Problem 1.1.-1 : Given a function f in an unlimited region :

$$f = 10x + 12x^2 + 12y^2 - 3x^3 - 9x^2y - 9xy^2 - 3y^3 .$$

Find out the extrema.

This problem has to illustrate the application of the necessary and the sufficient conditions for extrema of functions.

At first, we see that f is not bounded :

If $y = 0$ and x tends to infinity we see :

$$x \longrightarrow +\infty \quad : \quad f \longrightarrow -\infty ,$$

$$x \longrightarrow -\infty \quad : \quad f \longrightarrow +\infty .$$

Thus, there is no absolute extremum.

To find out relative extrema, we have to use the derivatives

$$f_{,x} \equiv \frac{\partial f}{\partial x} \ ; \quad f_{,y} \equiv \frac{\partial f}{\partial y} \ ; \quad f_{,xy} \equiv \frac{\partial^2 f}{\partial x \partial y} \ ;$$

$$f_{,xx} \equiv \frac{\partial^2 f}{\partial x^2} \ ; \quad f_{,yy} = \frac{\partial^2 f}{\partial y^2} \ :$$

$$f_{,x} = 10 + 24x - 9x^2 - 18xy - 9y^2 ,$$

$$f_{,y} = 24y - 9x^2 - 18xy - 9y^2 ,$$

$$f_{,xx} = 24 - 18(x+y) ,$$

$$f_{,xy} = -18(x+y) ,$$

$$f_{,yy} = 24 - 18(x+y) .$$

Necessary condition for an extremum of a continually differentiable function is stationarity :

$$f_{,x} = 0 \ , \quad f_{,y} = 0 .$$

This yields the two points :

$$\tilde{x}_1 = \frac{5}{8} \ ; \quad \tilde{y}_1 = 15/24 \ ; \quad \tilde{f}_1 = 725/72 ,$$

$$\tilde{x}_2 = -\frac{3}{8} \ ; \quad \tilde{y}_2 = 1/24 \ ; \quad \tilde{f}_2 = -139/72 .$$

But we do not know whether these points represent relative extrema. We examine the matrix

$$\frac{\partial^2 f}{\partial x^2} \equiv \begin{bmatrix} f_{,xx} & f_{,xy} \\ f_{,xy} & f_{,yy} \end{bmatrix}, \quad \text{taken at} \quad x = \tilde{x}, y = \tilde{y}, f = \tilde{f}.$$

If it is positive or negative definite we have a minimum or a maximum resp., if it is positive or negative semidefinite we possibly may have a minimum or maximum resp., but then we cannot be sure. If $\partial^2 f / \partial x^2$ is not semidefinite we have no extremum but a saddle-point. Applying this we see :

point 1 $$\left(\frac{\partial^2 f}{\partial x^2}\right)_1 = \begin{bmatrix} -6 & -30 \\ -30 & -6 \end{bmatrix} \equiv A_1 .$$

We check the definiteness by a direct method. We introduce the vector $\eta = (\alpha \ \beta)$, then $g = \eta A < \eta >$ is examined :

$$g_1 = \eta A_1 < \eta > = -6(\alpha^2 + \beta^2) - 60\alpha\beta .$$

We see :

$$g_1 = -72 \quad \text{if} \quad \alpha = \beta = 1,$$

$$g_1 = +48 \quad \text{if} \quad \alpha = -\beta = 1.$$

A_1 is not definite or semidefinite: point 1 is a saddle-point.

point 2 $$\left(\frac{\partial^2 f}{\partial x^2}\right)_2 = \begin{bmatrix} 30 & 6 \\ 6 & 30 \end{bmatrix} \equiv A_2 .$$

This yields

$$g_2 = \eta A_2 < \eta > = 24(\alpha^2+\beta^2) + 6(\alpha+\beta)^2 > 0 \text{ if } \alpha \neq 0 \text{ or } \beta \neq 0 .$$

Hence, A_2 is positive definite, point 2 represents a relative minimum.

The function f has only one minimum and no maximum. This is possible if it has the form which is sketched in Fig. 1. 1. -1.

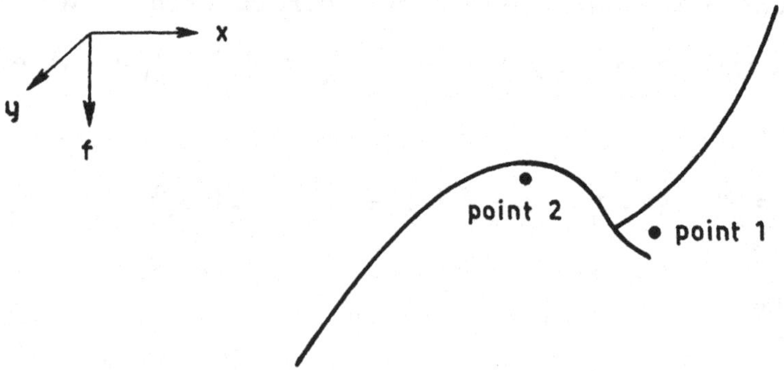

Fig. 1.1-1

Problem 1.1-2 : Given the function $f = 3x^2 + 4y^2 + z^2$ declaired in the region G where $g \equiv x^2 + y^2 + z^2 - 1 \le 0$ (unit sphere), calculate the extrema, also boundary-extrema.

This problem has to show the curious effect that extrema can be lost if we are not careful enough when calculating extrema on boundaries.

First we try to find out extrema in the interior of the region G :

$$f_{,x} = 6x \; ; \quad f_{,y} = 8y \; ; \quad f_{,z} = 2z \; .$$

Hence, $f_{,x_i} = 0$ leads to $\tilde{x} = \tilde{y} = \tilde{z} = \tilde{f} = 0$.

Because of $f \ge 0$ this must be a minimum. There is no second extremum in the interior.

Inside of G, f is bounded. So there must be a maximum on the boundary.

The boundary is described by $g = 0$. Therefore, $z^2 = 1 - x^2 - y^2$ can be put into f instead of z^2. So we get the function f as a function $\overset{B}{f} = \overset{B}{f}(x,y)$ which is valid on the boundary :

$$\overset{B}{f} = 1 + 2x^2 + 3y^2$$

$\overset{B}{f}_{,x} = \overset{B}{f}_{,y} = 0$ yields $x = y = 0$, $f = \overset{B}{f} = 1$ which is a minimum of $\overset{B}{f}$ and, therefore, cannot be a maximum of f.

Two questions arise now :

1. Is $x = y = 1$, $f = 1$, $z = \pm 1$ a minimum of the function f in G?

The allowed region is given by $g \leq 0$. Thus, the gradient $\mathrm{grad}\ g$ represents a vector directed towards the outside of G. Then

$$(\mathrm{grad}\ f) \cdot (\mathrm{grad}\ g) \lesseqgtr 0$$

is a necessary condition for a minimum or a maximum on the boundary.

$(\mathrm{grad}\ f) \cdot (\mathrm{grad}\ g) \gtrless 0$ at a special point x_i is sufficient, if there is a minimum or maximum, resp., on the boundary.

$$(\mathrm{grad}\ g) \cdot (\mathrm{grad}\ f) = 12\,x^2 + 16\,y^2 + 4\,z^2 > 0$$

shows that we can find out only maxima on the boundary.

2. We have found out only two stationary points on the boundary. They did not represent maxima. But there must be a maximum. It seems to be lost. What is the reason?

We have lost the maximum because of the following mistake which we made : When establishing the function $\overset{B}{f}$, we did not notice that z^2 must be positive. This would have lead to the new boundary condition for $\overset{B}{f}$:

$$\overset{B}{g} \equiv x^2 + y^2 - 1 \leq 0 .$$

We see : the elimination of variables, especially of squared ones, may be dangerous.

The other way for the calculation of the stationarities of f on the boundary is the use of LAGRANGIAN multipliers. The problem : $f \Longrightarrow$ stationary under the side-condition $g = 0$, can be expressed as

$$h(x_i, \lambda) = 3x^2 + 4y^2 + z^2 + $$
$$+ \lambda (x^2 + y^2 + z^2 - 1) \Longrightarrow \quad \text{stationary}$$

This method yields each stationary point on the boundary which, then, can be checked, whether it is an extremum .

$h \Longrightarrow$ stationary yields the conditions :

$$6\tilde{x} + 2\tilde{\lambda}\tilde{x} = 0 \quad \text{or} \quad (6 + 2\tilde{\lambda})\tilde{x} = 0 \qquad (1)$$

$$8\tilde{y} + 2\tilde{\lambda}\tilde{y} = 0 \quad \text{or} \quad (8 + 2\tilde{\lambda})\tilde{y} = 0 \qquad (2)$$

$$2\tilde{z} + 2\tilde{\lambda}\tilde{z} = 0 \quad \text{or} \quad (2 + 2\tilde{\lambda})\tilde{z} = 0 \qquad (3)$$

$$\tilde{x}^2 + \tilde{y}^2 + \tilde{z}^2 = 1 . \qquad (4)$$

Eq. (1) postulates $\tilde{x} = 0$ or $\tilde{\lambda} = -3$,
Eq. (2) yields $\tilde{y} = 0$ or $\tilde{\lambda} = -2$, and
Eq. (3) leads to $\tilde{z} = 0$ or $\tilde{\lambda} = -1$.

$\tilde{x} \neq 0$ is possible if $\tilde{\lambda} = -3$ but then we must have $\tilde{y} = \tilde{z} = 0$, $\tilde{x} = \pm 1$ (from eq. (4)). In the same way we get :

$\tilde{y} \neq 0$ leads to $\tilde{x} = \tilde{z} = 0$, $\tilde{y} = \pm 1$,

$\tilde{z} \neq 0$ brings out $\tilde{x} = \tilde{y} = 0$, $\tilde{z} = \pm 1$.

These are points of stationarity on the boundary.

The examination of their extremum properties shows that the points

$$\tilde{x} = \tilde{z} = 0 , \quad \tilde{y} = \pm 1 , \quad \tilde{f} = 4$$

represent the (absolute) maxima of f in G .

Problem 1.1-3 : We have a given plate of sheet metal and we want to produce with it a fixed number of tin-boxes. Calculate the optimum relation between the height h and the radius r of the tin boxes, if the volume of the boxes is to be maximized.

This simple problem was used as an additional problem to demonstrate the advantage of LAGRANGE-multipliers.

The volume

$$V = \pi r^2 h$$

(r = radius, h = height) has to be a maximum. On the other hand, the surface area of the plate per box

$$S = 2\pi r h + \gamma \pi r^2 \qquad (\gamma \geq 2)$$

is a given value. The quantity γ is introduced, because differ ent cases will be examined : ideally no falling-off leads to $\gamma = 2$, falling-off as sketched in Fig. 1.1-2 belongs to $\gamma = 2.20$, realistic values of γ may be 2.30 to 2.60.

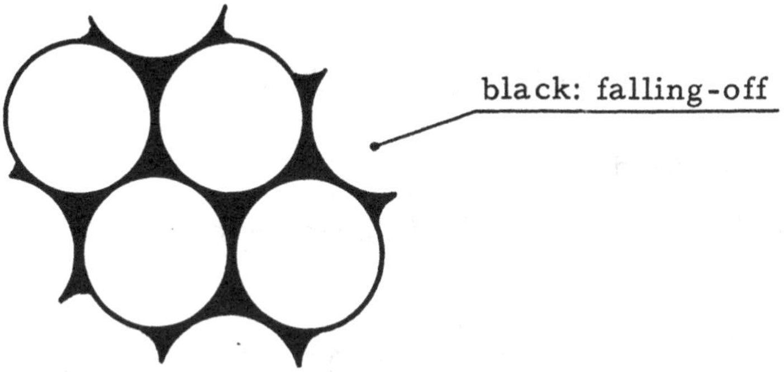

Fig. 1.1-2

The problem is fixed now :

$$V = \pi r^2 h \Longrightarrow \text{maximum}$$

under the side-condition

$$g = 2\pi r h + \gamma \pi r^2 - S = 0.$$

This leads to $H(r, h, \lambda)$ being

$$H(r, h, \lambda) = \pi r^2 h -$$

$$- \lambda(2\pi r h + \gamma \pi r^2 - S) \Longrightarrow \quad \text{stationary}$$

Hence, we get

(1) $$\tilde{r}\tilde{h} - \tilde{\lambda}\tilde{h} - \tilde{\lambda}\gamma\tilde{r} = 0 ,$$

(2) $$\tilde{r}^2 - \tilde{\lambda} 2\tilde{r} = 0 ,$$

(3) $$2\pi\tilde{r}\tilde{h} + \gamma\pi\tilde{r}^2 - S = 0.$$

Eq. (2) yields : $\tilde{\lambda} = \frac{1}{2}\tilde{r}$.

Putting this $\tilde{\lambda}$ into eq. (1), we reach

$$\tilde{r}\tilde{h} - \frac{1}{2}\tilde{r}\tilde{h} - \frac{1}{2}\gamma\tilde{r}^2 = \frac{1}{2}(\tilde{r}\tilde{h} - \gamma\tilde{r}^2) = \frac{1}{2}\tilde{r}(\tilde{h} - \gamma\tilde{r}) = 0$$

or : $\tilde{r} = 0$ or $\tilde{h} = \gamma\tilde{r}$.

$\tilde{r} = 0$ cannot be a maximum of V ($V = 0$) . Therefore, $\tilde{h} : \tilde{r} = \gamma$ must be the desired result. The ideal tin-box has the form sketched in Fig. 1.1-3 (γ = 2 5). Eq. (3) can be used for the determination of $\tilde{r}$, $\tilde{h}$ as functions of S :

$$2\gamma\pi\tilde{r}^2 + \gamma\pi\tilde{r}^2 = S ,$$

$$\tilde{r} = \sqrt{\frac{S}{3\gamma\pi}} ; \quad \tilde{h} = \sqrt{\frac{\gamma S}{3\pi}}$$

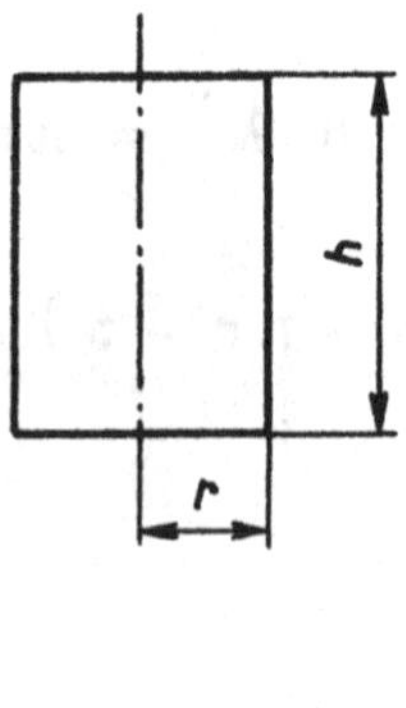

Fig. 1.1-3

1.2. Linear programming (no correspondence to the lectures)

In a lot of problems where a minimum or a maximum has to be calculated, the equations describing the boundaries and the functions which have to be optimized, are linear in their variables. Then, the method of "Linear Programming" can be applied. We will derive the theory by use of a simple example. Later on, this method will be applied for the calculation of the load-carrying capacity.

Problem 1.2-1 : A farmer has 100 ha (German unit of measurement, 1 ha = 10,000 m^2) grounds on which he wants to cultivate four types of fruits (I to IV) in order to reach maximal profit. For this purpose, he has to use different means which are restricted : his capital and the working-time are not infinite. Further on, he has to use two machines A and B which he has to lend. This is possible for restricted times only.

We assume that costs and times depend linearly on the area which is cultivated with a special fruit. Also the profit (where all costs are subtracted already) is to be a linear function of the parts of the grounds cultivated with the different fruits. Then, the theory of linear programming can be applied.

Two problems will be handled :

a) Only two fruits (I and IV) are taken into account. Then, the two distributions of the profit α) and β) will be compared (cf. Table 1.2-1).

b) Four fruits are possibly cultivated. The profit-distribution α) is valid.

The values which are necessary for the calculation are printed in Table 1.2-1.

Table 1.2-1

means / fruits	money [lire/ha]	time [days/ha] for work	machA	machB	profit [Lire/ha] case α)	case β)
I	20,000	1	0	1/2	24,000	12,000
II	40,000	2	1	2	48,000	-
III	20,000	3	0	0	36,000	-
IV	30,000	4	1	0	54,000	60,000
restrictions for the sums:	[Lire] 2,750,000	230	[days] 50	75		

a) Two fruits (I and IV), profit given by α) and β)

The restrictions of the means lead to necessary conditions for every solution, given in Table 1.2-2 ($x = x_I$ = number of ha's cultivated by fruit I, $y = x_{IV}$).

Table 1.2-2

Number of condition	means	inequality	last possible value of y, if $x=0$	x, if $y=0$
1	ground	$x+y \leq 100$	100	100
2	time:work	$x+4y \leq 230$	57.5	230
3	capital	$2x+3y \leq 275$	91.667	137.5
4	time:mach. A	$y \leq 50$	50	∞
5	time:mach. B	$x \leq 150$	∞	150
6	trivial	$x \geq 0$	-	-
7		$y \geq 0$	-	-

Only that part of the x,y - plane where each of these seven relations is satisfied, is admissible. It is the (weakly convex) region I, II, III, IV, V of Fig. 1.2-1, whose corners I to IV are presented in Table 1.2-3.

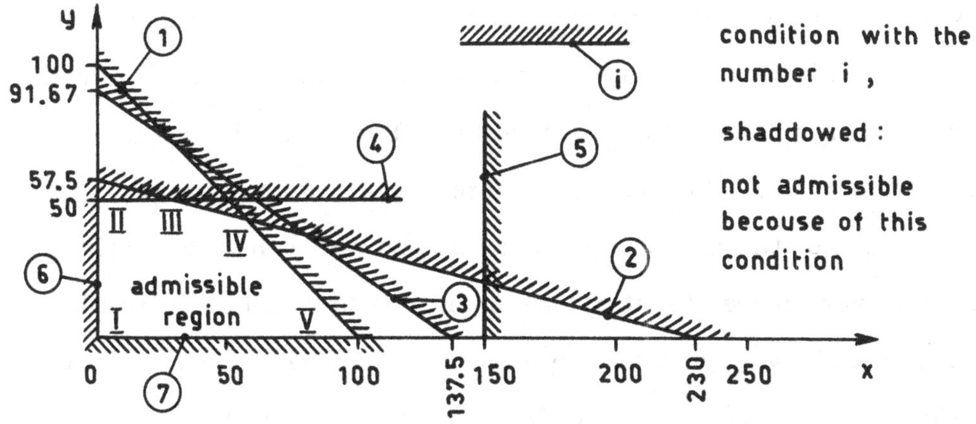

Fig. 1.2-1

As we see from Fig. 1.2-1, conditions 3 and 5 do not matter.

Table 1.2-3

point-number	x	y	profit α [Lire]	profit β [Lire]
I	0	0	0	0
II	0	50	2, 700, 000	3, 000, 000
III	30	50	3, 420, 000	3, 360, 000
IV	56. 667	43. 333	3, 700, 000	3, 380, 000
V	100	0	2, 400, 000	1, 200, 000

The profit is given by an equation $P = Ax + By + P_0$. Lines of constant profit (P = const.) are straight lines, the lines belonging to different values of P are parallel to each other. We can define a profit-gradient $\underline{g} = (A\ B)$ which is a vector normal to the lines of constant profit. $\underline{g}$ is constant.

Only if $\underline{g}$ vanishes, it is possible that the optimum is reached inside the admissible region. But, then P itself is constant. In any other case, the optimum can only be given at some point of the boundary. It is easy to show that the optimum must be reached at least at one corner (I to V in our example) of the boundary : As far as $\underline{g}$ has a not-vanishing component in the direction of a straight part of the boundary line, one corner belonging to this straight line gives the highest profit on this part of the boundary. If $\underline{g}$ is normal to the straight line the profit is constant on this line, then, every point of this line can be optimal, especially one corner, belonging to this line. So we can state :

<u>The optimum is reached in at least one corner of the admissible region.</u>

With this theorem, the optimum can be seen from a table like Table 1.2-3. For our example we find out :

<u>case α)</u> optimal point is point IV : $x = 56.67$; $y = 43.44$:

$P = 3,700,000$ Lire.

<u>case β)</u> optimal point is point III : $x = 30$; $y = 50$:

$P = 3,360,000$ Lire.

The second result is a rather strange one, because not the complete ground is cultivated.

<u>b) Four fruits, profit according to case α</u>

When taking into account four fruits, we get nine restrictions of the kind

$$a_I x_I + a_{II} x_{II} + a_{III} x_{III} + a_{IV} x_{IV} \lesseqgtr b .$$

The coefficients a_i and b of these conditions are given at Table 1.2-4.

Table 1.2-4

Condition number	reason	a_I	a_{II}	a_{III}	a_{IV}	sign	b
1	ground	1	1	1	1	$\leq$	100
2	work-time	1	2	3	4	$\leq$	230
3	costs	2	4	2	3	$\leq$	275
4	mach. A	0	1	0	1	$\leq$	50
5	mach. B	1	4	0	0	$\leq$	150
6	trivial	1	0	0	0	$\geq$	0
7	trivial	0	1	0	0	$\geq$	0
8	trivial	0	0	1	0	$\geq$	0
9	trivial	0	0	0	1	$\geq$	0

In future, we will speak of "eq. (i)" if we want to express that we take the condition (i) with an equality sign instead of the inequality sign, whereas "condition (i)" means the original condition.

Each of these conditions defines a hyperplane as a boundary between admissible and non-admissible half-spaces in an n-space ($n = 4$ in our example). These half-spaces are convex. The conditions (1) to (9) must be satisfied. Therefore, the admissible region is the intersection of these half-spaces. The boundary of this intersection consists of hyperplanes of different order. If $\underline{q}$ is normal to such a hyperplane, each corner belonging to it, represents an optimum point ($\underline{q}$ = gradient of profit). If, on the other hand, $\underline{q}$ has a component in the considered boundary-hyperplane, a point of the boundary of this hyperplane part of the boundary of the admissible region must be better than all other points of this hyperplane. This partly optimal point belongs to a boundary-hyperplane with a reduced number of degrees of freedom. Here, the same considerations can be done, until the corner points of the admissible region are reached. So we see :

<u>The optimum is reached in at least one corner of the admissible region.</u>

One exception has to be mentioned : Possibly, the admissible region is not closed, but contains infinite values of the coordinates. We assume that the optimum will not belong to points with infinite coordinates. Then, our theorem of optima in corners of the admissible region remains valid.

We have to discuss how these corners can be determined.

At every part of the boundary of the admissible region, at least one equation ((1) to (9)) is satisfied. At a corner of the admissible region in a n -dimensional problem, at least n equations (i) are fulfilled. This shows the way of calculation of these corner points :

n equations are tried to be solved. If they define a point in the n-space, we have to check whether the other conditions are satisfied. Otherwise, the computed point is no corner of the admissible region. Possibly, n equations do not define a point. Then, the determinant of the coefficient matrix vanishes. In such a case, it must be checked whether the n equations are contradictionless. If contradictions occur no point can satisfy these n equations, so they cannot define any corner. Otherwise, a $(n+1)$th equation can be added. We derive two solution methods from these considerations :

I. Calculation of each corner of the admissible region.

If we have m conditions and n variables, we handle each of the possible $\binom{m}{n}$ combinations of n conditions. In our example, we have $\binom{m}{n} = \binom{9}{4} = \frac{9\cdot 8\cdot 7\cdot 6}{1\cdot 2\cdot 3\cdot 4} = 126$.

Computing each corner in this way, we get 19 corners which are present in Table 1.2-5 (see next page).

Knowing the corners, we compute the profit at these corners. The maximum of these 19 values is the optimum. So we see : point 1 leads to the optimal profit of 3,930,000 Lire.

II) Iteration method.

This method is much quicker than the other procedure, especially, if only one vector $\underline{g}$ has to be treated.

For the interpretation, we define "neighbouring-points":

Table 1.2-5

point number	x_{I}	x_{II}	x_{III}	x_{IV}	satisfied equations 1	2	3	4	5	6	7	8	9	profit [Lire]
1	35	25	15	25	x	x	x	x						3 930 000
2	26	31	30	13	x	x	x		x					3 894 000
3	50	10	0	40	x	x		x				x		3 840 000
4	18.57	32.86	48.57	0	x	x			x				x	3 770 000
5	56.67	0	0	43.33	x	x					x	x		3 700 000
6	35	0	65	0	x	x					x		x	3 180 000
7	50	25	0	25	x		x	x	x			x		3 750 000
8	83.33	16.67	0	0	x				x			x	x	2 800 000
9	100	0	0	0	x						x	x	x	2 400 000
10	23.33	31.67	23.33	18.33		x	x	x	x					3'910'000
11	0	37.5	35	12.5		x		x	x	x				2 935 000
12	0	0	10	50		x		x		x	x			3 060 000
13	30	0	0	50		x		x			x	x		3 420 000
14	0	37.5	51.67	0		x			x	x			x	3 660 000
15	0	0	76.67	0		x				x	x		x	2 760 000
16	0	37.5	0	12.5				x	x	x		x		2 575 000
17	0	0	0	50				x		x	x	x		2 700 000
18	0	37.5	0	0					x	x		x	x	1 800 000
19	0	0	0	0						x	x	x	x	0

At a corner at least n equations (i) are satisfied. $n-1$ of these equations (if more than n equations are satisfied at the corner, this number can be higher) define a straight line in the n-space. Each of these straight lines belongs to the boundary of the admissible region at two corners if n equations are satisfied at the starting-corner, otherwise, we can only say that it can contain two corners.

The other corners on these straight lines will be called "neighbouring-points".

Now we state :

If there is one corner where the profit (or any other function which has to be maximized) is not less than at any of its neighbouring points, this point is an optimum point.

Proof : The admissible region is convex, because it is the intersection of convex half-spaces. A plane normal to $\underline{g}$ crossing the considered point of (local) optimum, does not intersect the admissible region in the neighbourhood of this point. Because of the convexity, it then cannot intersect the admissible region at any other point. So the complete admissible region lies on one side of this plane which, on the other hand, is the surface of equal profit. Therefore, the profit of any other admissible point cannot be higher, because at least one of its neighbouring points has a lower profit.

We derive an iteration method from these considerations. First, there must be any known corner-point. Then, it can be checked, whether any neighbouring point of this corner leads to a higher profit. If one does, this point is used as the starting point of the next step. If not, the considered point represents the optimum. This method must converge, because the optimized function is bettered at each step and the number of corners is limited.

At our example, we can imagine that we start our calculations at point 9 (Table 1.2-5) where the profit is 2,400,000 Lire. Here the equations (1), (7), (8), and (9) are satisfied. The combinations (1), (7), (8) ‖ (1), (7), (9) ‖ (1), (8), (9) ‖ (7), (8), (9) lead to neighbouring points. Let us use (1), (8), (9). Then, after having checked 4 combinations, we reach point 8 with eqs. (1), (5), (8), (9) and with a profit of 2,800,000 Lire. Now we start at this point. Eqs. (5), (8), and (9) lead after 4 combinations to point 18 with a profit of 1,800,000 Lire. But eqs. (1), (5), (9) lead to point 4 with a profit of 3,770,000 Lire.

In this way we reached - in a real calculation - the optimum point 1 after having checked 18 combinations of 4 equations, 6 further combinations were necessary to show that it was the optimal point. So we used 24 combinations instead of 126 at the first-mentioned method.

1.3. Rayleigh-quotient and eigenvalues of matrices

(cf. sect. 1.2 of the lecture-notes).

In this chapter we use the matrices A and B

$$A = \begin{bmatrix} 103 & -18 & -2 & 30 & -9 \\ -18 & 116 & -52 & 4 & 30 \\ -2 & -52 & 60 & -52 & -2 \\ 30 & 4 & -52 & 116 & -18 \\ -9 & 30 & -2 & -18 & 103 \end{bmatrix}, \quad B = \begin{bmatrix} 19 & 6 & -2 & 6 & 3 \\ 6 & 20 & -4 & 4 & 6 \\ 2 & -4 & 12 & -4 & -2 \\ 6 & 4 & -4 & 20 & 6 \\ 3 & 6 & -2 & 6 & 19 \end{bmatrix}$$

For our considerations, the symmetry of A and B with respect to the main diagonal is assumed, whereas their symmetry with respect to the other diagonal is of no interest.

Problem 1.3-1 : Based on the minimum property of the lowest eigenvalue r^1 of the problem

$$\left[A - r B\right] \langle \tilde{x} \rangle = 0 ,$$

develop an iteration method for this value r^1 and for the eigenmode $\tilde{x}^1$ belonging to it, which uses two possible modes at each step of the iteration.

The method based on this theory will not be the best one. But it is a possible iteration method which converges under arbitrary conditions as far as $B \gtrsim 0$ is given.

The lowest eigenvalue r^1 of the problem

$$(A - rB) < \tilde{x} > = 0 \tag{1}$$

is connected with the minimum of the quotient

$$r(x) = \frac{x A < x >}{x B < x >} . \tag{2}$$

The iteration method provides to search for this minimum by a stepwise minimization of this quotient. At each step, we find out the minimum of $r(x)$ in a 2-dimensional sub-space where x is a linear combination of two modes x^1 and x^2.

Thus, x has the form

$$x = y \cdot q \qquad \text{where} \qquad q = \begin{bmatrix} x^1 \\ \cdots \\ x^2 \end{bmatrix} \tag{3}$$

is a matrix with two lines. Within the sub-space $\min(r)$ is given by

$$r_{min}\Big|_{x = y \cdot q} = \min \left(\frac{x A < x >}{x B < x >} \right)\Bigg|_{x = y \cdot q} = \min \left(\frac{y q A < q > < y >}{y q B < q > < y >} \right),$$

which can be written as

$$\left.\begin{aligned} & r_{min}\Big|_{x = y \cdot q} = \min \left(\frac{y \hat{A} < y >}{y \hat{B} < y >} \right), \\ \text{where} \qquad & \hat{A} = q A < q > , \ \hat{B} = q B < q > . \end{aligned}\right\} \tag{4}$$

Then, $r_{min}\Big|_{x = y \cdot q}$ is the minimal solution $\hat{r}$ of the problem

(5) $$\left(\hat{A} - \hat{r}\hat{B}\right)\langle \tilde{y} \rangle = 0.$$

This problem can always be solved. It yields a vector $\tilde{y}$ and afterwards $x^3 = \tilde{y}q$. This x^3 is taken as x^1 for the next step of the iteration. The solution $\hat{r}$ belongs to

(6) $$\det\left(\hat{A} - \hat{r}\hat{B}\right) = 0.$$

The solution of eq. (6) is easier than in a general case if $\hat{B}$ or $\hat{A}$ have a diagonal form. But it is not advantageous if $\hat{A}$ <u>and</u> $\hat{B}$ are diagonal matrices as we will see later.

The coordinates of $\hat{A}$ and $\hat{B}$ are given by

(7) $$\hat{A}_{ij} = x^i A \langle x^j \rangle \quad ; \quad \hat{B}_{ij} = x^i B \langle x^j \rangle .$$

The diagonal form of $\hat{B}$ then postulates

(8) $$x^1 B \langle x^2 \rangle = 0.$$

This equation can always be satisfied. Under these conditions eq. (6) yields

(9) $$\hat{r}_{min} = \frac{\hat{A}_{11}\hat{B}_{22} - \hat{A}_{22}\hat{B}_{11}}{2\hat{B}_{11}\hat{B}_{22}} - \sqrt{\left(\frac{\hat{A}_{11}\hat{B}_{22} - \hat{A}_{22}\hat{B}_{11}}{2\hat{B}_{11}\hat{B}_{22}}\right)^2 + \frac{\hat{A}_{12}^2}{\hat{B}_{11}\hat{B}_{22}}} .$$

In the lecture course, B was restricted to be definite. Thus, $\hat{B}_{11}$ and $\hat{B}_{22}$ have equal sign. Therefore, $\hat{A}_{12}^2/(\hat{B}_{11}\hat{B}_{22})$ is positive if $\hat{A}_{12} \neq 0$.

For $\hat{A}_{12} = 0$, eq. (9) yields

$$r_{min} = \min \left(r(x^1), r(x^2) \right).$$

Then, the iteration does not work : x^3 is identical with x^1 or x^2. We see,that $\hat{A}_{12} \neq 0$ is necessary. This means : eq. (8) is an advantageous restriction, but a similar expression for A instead of B should not be satisfied simultaneously.

If $\hat{A}_{12} \neq 0$ $\hat{r}_{min}$ is lower than $r(x^1)$ or $r(x^2)$. Thus, our method is a possible iteration method.

As an example we use the matrices A and B. Starting from $x^1 = (0 \quad 1 \quad 1 \quad 1 \quad 0)$, eq. (8) states for $x^2 =$ $= (a \; b \; c \; d \; e)$.

$10a + 20b + 4c + 20d + 10e = 0$. This is, for example, satisfied if x^2 is given by

$$x^2 = (2 \quad -1 \quad 0 \quad -1 \quad 2)$$

or

$$\bar{x}^2 = (1 \quad 0 \quad 0 \quad 0 \quad 1).$$

If we take $\bar{x}^2$, we get

$$\hat{A} = \begin{bmatrix} 92 & 0 \\ 0 & 224 \end{bmatrix}, \quad \hat{B} = \begin{bmatrix} 44 & 0 \\ 0 & 32 \end{bmatrix}, \quad \hat{r}_{min} = r(x^1) = 2{,}0909.$$

This is no iteration, but x^2 leads to

$$\hat{A} = \begin{bmatrix} 92 & -96 \\ -96 & 896 \end{bmatrix}, \quad \hat{B} = \begin{bmatrix} 44 & 0 \\ 0 & 128 \end{bmatrix}, \quad \hat{r}_{min} = 1{,}778 < 2{,}0909.$$

For comparison we calculate $r(x^2)$:

$$r(x^2) = \frac{896}{128} = 7.$$

The iteration works, but the method is not the best one, as we mentioned above.

Problem 1. 3-2 : Use the properties of the RAYLEIGH-quotient for the reduction of the matrices A and B in cases where several relations between the $\tilde{x}_i$ are known.

Stationarity of r defined by eq. (2) belongs to each solution r of eq. (1) (problem 1. 3-1). This stationarity must also be given in any sub-space. On the other hand, in this subspace, a solution of (1) can be found to each stationarity of r according to eq. (2). Therefore, if we know that a solution with certain relations between the values $\tilde{x}_i$ exists we will find out this solution in the sub-space of those x^i which satisfy these relations. Thus, we can use eqs. (3), (4), (6), and (7) with

$$q = \begin{bmatrix} x^1 \\ \vdots \\ x^i \\ \vdots \\ x^m \end{bmatrix},$$

where $m = n - p$ (p being the number of given linear relations) and x^i being linearly independent. Advantageously, the x^i are orthogonal to each other with respect to B. Then, they satisfy

$$x^i B < x^j > = a \delta^{ij} \; , \quad a \neq 0 \; , \quad \delta^{ij} = \text{KRONECKER-symbol}$$

The modes $x = \tilde{y} q$ calculated by this method must contain the desired solution. A similar reduction method will later be used for approximate solutions for continuous systems. In our example, we may state that at least one solution $\tilde{x}$ should have the form

$$x = (a \; b \; c \; b \; a).$$

Then, $x^i B < x^j > = a_{ij} \delta^{ij}$ is satisfied by

$$x^1 = (0\;1\;1\;1\;0) \; ; \; x^2 = (2\;-1\;0\;-1\;2); \; x^3 = (0\;1\;-10\;1\;0) \; ,$$

which yields

$$q = \begin{bmatrix} 0 & 1 & 1 & 1 & 0 \\ 2 & -1 & 0 & -1 & 2 \\ 0 & 1 & -10 & 1 & 0 \end{bmatrix},$$

$$\hat{A} = \begin{bmatrix} 92 & -96 & 576 \\ -96 & 896 & -1152 \\ 576 & -1152 & 8320 \end{bmatrix}, \qquad \hat{B} = \begin{bmatrix} 44 & 0 & 0 \\ 0 & 128 & 0 \\ 0 & 0 & 1408 \end{bmatrix}.$$

$\det(\hat{A} - \hat{r}\hat{B}) = 0$ then leads to

$$\hat{r}^3 - 15\,\hat{r}^2 + 54\,\hat{r} - 40 = 0.$$

The roots of this equation are $\hat{r}^1 = 1$, $\hat{r}^2 = 4$, $\hat{r}^3 = 10$. Inserting these values $\hat{r}^i$ into the equation $\det(A - rB) = 0$, we see that the $\hat{r}^i$ are solutions of our original problem. The corresponding eigenmodes are :

$$\tilde{x}^1 = (0\ \ 1\ \ 2\ \ 1\ \ 0);\ \ \tilde{x}^2 = (1\ \ 0\ \ {-1}\ \ 0\ \ 1);\ \ \tilde{x}^3 = (1\ \ {-1}\ \ 1\ \ {-1}\ \ 1).$$

In the same way the assumption $\tilde{x} = (a\ b\ 0\ {-b}\ {-a})$ leads to $\hat{r}^1 = 4$, $\hat{r}^2 = 10$.

These eigenvalues of this reduced system are eigenvalues of the original system, too. So we get

$$r^4 = 4\ ,\ r^5 = 10 \quad \text{with}$$

$$\tilde{x}^4 = (1\ \ 1\ \ 0\ \ {-1}\ \ {-1});\ \ \tilde{x}^5 = (1\ \ {-1}\ \ 0\ \ 1\ \ {-1}).$$

2. PRINCIPLES BASED ON THE VIRTUAL WORK THEOREM

2.1. The virtual work theorem itself (cf. sect. 2.2 of the lecture-notes).

The method of virtual work in statics is well-known and will be applied later in more complicated problems. Therefore, only a very simple example is mentioned presently.

Problem 2.1-1 : Find out the bending moment M_b at point A of the framework sketched in Fig. 2.1-1 which is loaded by the constant load per unit length q

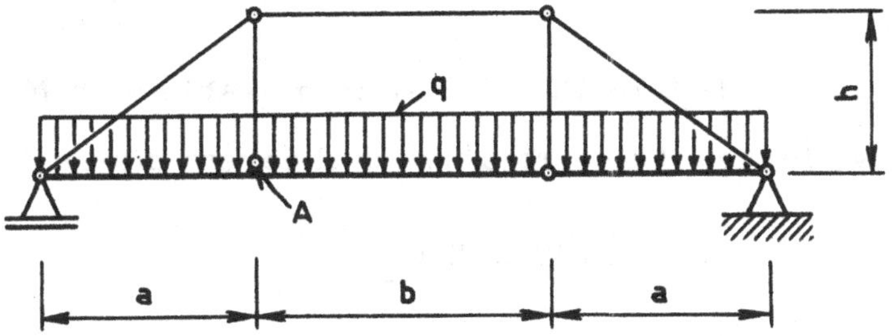

Fig. 2.1-1

For the application of virtual work theorems, we must have movable systems or make rigid systems movable by introducing hinges, etc. The method is, to do this in such a way that the desired values can be computed as easy as possible. So, in our problem, we introduce a hinge at A (cf. Fig. 2.1-2).

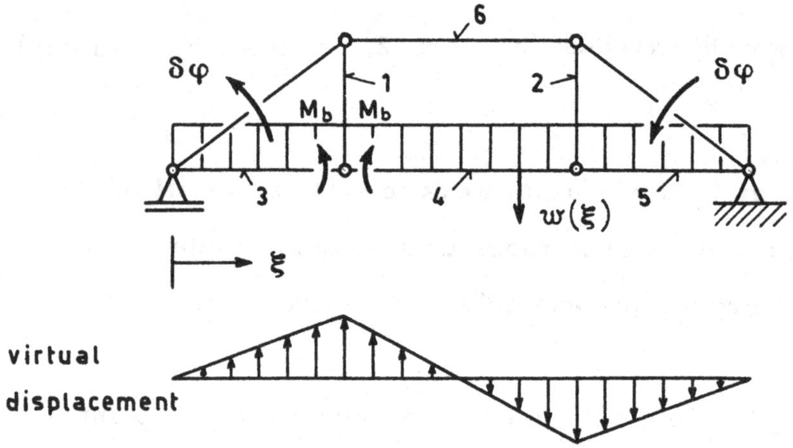

Fig. 2.1-2

The system of Fig. 2.1-2 is movable and M_b is now, an effective force. The virtual work theorem states

$$\delta' W = \delta' W_s = 0$$

(cf. lecture-notes), where $\delta' w_s$ is given by

$$\delta' W_s = \int q\, \delta' w\, d\xi - M_b \delta' w_3' + M_b \delta' w_4' = 0$$

$(w' \equiv \partial w / \partial \xi)$. First we have to find out $\delta' w(\xi)$ as a function of ξ and any parameter. The parallelism of 1, 2, 4, and 6 in Fig. 2.1-2 brings out that the left triangle moves parallel to the right triangle (cf. angle $\delta\varphi$). Within one bar (3,

4, or 5) the distribution of $\delta' w$ is linear. So we see easily that $\delta' w(\xi)$ must be printed like in the lower half of Fig. 1.2-2. For this distribution

$$\int q\, \delta' w(\xi)\, d\xi = q \int \delta' w(\xi)\, d\xi$$

(q = const.) vanishes. But the difference $\delta' w_4' - \delta' w_3'$ does not vanish. Therefore, M_b has to be zero :

$$\underline{\underline{M_b = 0.}}$$

2.2. Problems with holonomic constraints (cf. sect. 2. 2 of the lecture-notes).

In this section and in the next one the following principles will be compared and illustrated :

a) LAGRANGE equations of the 1st kind,

b) d'ALEMBERT's principle,

c) GAUSS-principle.

In each of these methods, the number of degrees of freedom is of interest only for the question how many constraints are needed. The number of introduced coordinates is not restricted. Two conditions have to be satisfied :

I) We have to introduce so many coordinates that every possible

α) effective force,

β) inertia force

can be described.

II) We can introduce maximally three coordinates to one body

α) 3 cartesian coordinates (point-masses in arbitrary motion)

β) 2 cartesian coordinates of the centre of mass and one angle (plane motion of rigid bodies).

If necessary, we have to introduce bodies with zero mass. This will be shown at our second problem.

Problem 2. 2-1 : Find out equations of motion for the problem sketched in Fig. 2. 2-1 by use of the three principles (a, b, c) mentioned above (no friction).

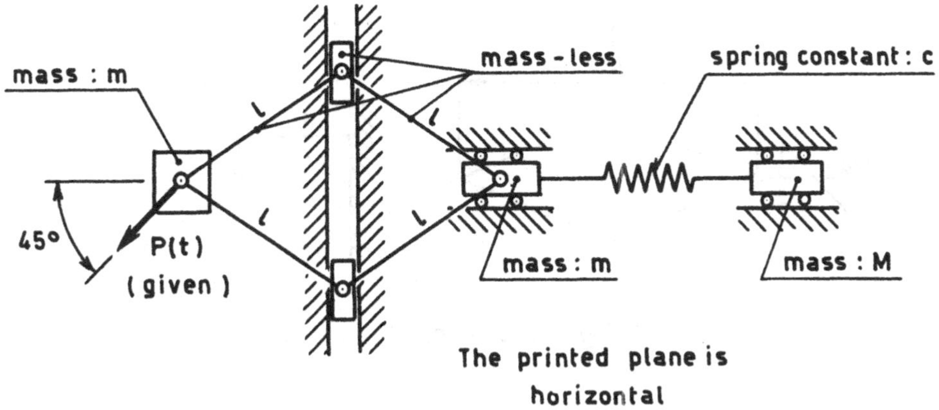

Fig. 2. 2-1

For the considered principles, the spring-force K must be treated to be an external effective force. Therefore, we have to introduce coordinates in the directions of the three forces printed in Fig. 2. 2-2a . Fig. 2. 2-2b shows the possible directions of inertia forces, which have to be described, too.

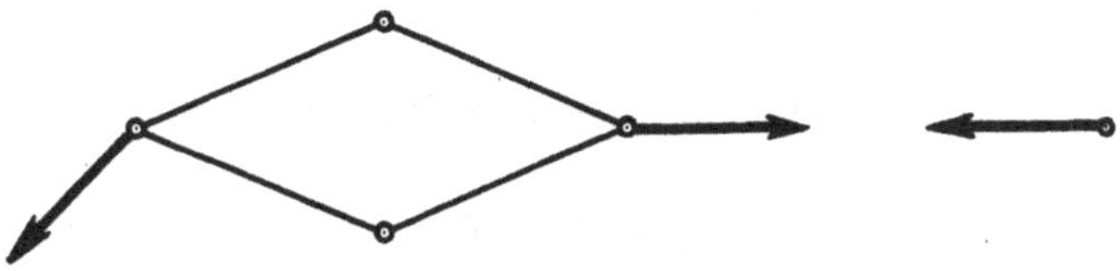

Fig. 2. 2-2 a)

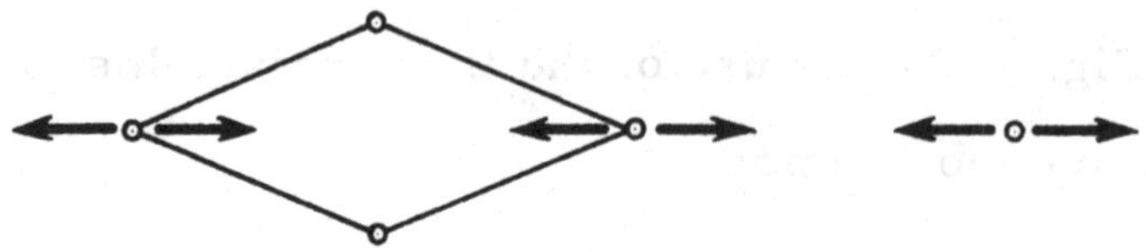

Fig. 2.2-2b)

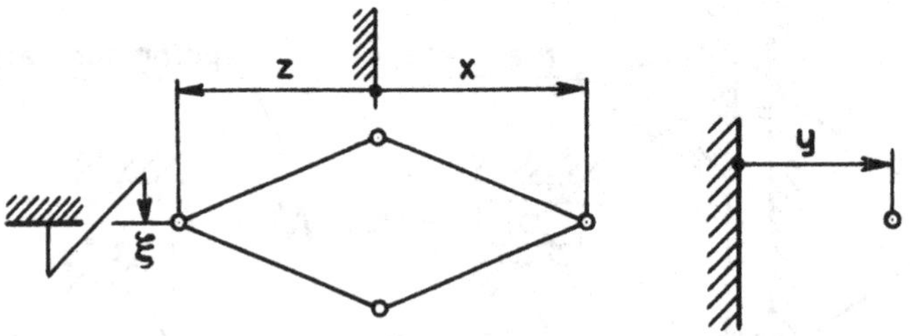

Fig. 2.2-2 c)

We see from Fig. 2.2 that the possibly interesting forces can be described by the coordinates printed in Fig. 2.2-2c. y has to be chosen in such a way that $K = 0$ if $x = y$. Thus, we get

$$K = c(y - x).$$

We have introduced four coordinates, but the system has only two degrees of freedom, so we must add two constraints :

(1) $$\varphi_1 \equiv z - x = 0,$$

(2) $$\varphi_2 \equiv \xi = 0.$$

We state that x and y will be the free coordinates. Now

we can build a table which shows what we have to use as x_j, m_j, P_j, φ_{ij} etc. (cf. lecture-notes).

Table 2. 2-1

x_j	m_j	P_j	principle a φ_{1j}	φ_{2j}	principle b $\delta' x_j (\delta' x, \delta' y)$	principle c $\delta'' \ddot{x}_j (\delta'' \ddot{x}, \delta'' \ddot{y})$
x	m	$c(y-x)$	-1	0	—	—
y	M	$c(x-y)$	0	0	—	—
z	m	$\frac{\sqrt{2}}{2} P$	1	0	$\delta' z = \delta' x$	$\delta'' \ddot{z} = \delta'' \ddot{x}$
ξ	m	$\frac{\sqrt{2}}{2} P$	0	1	$\delta' \xi = 0$	$\delta'' \ddot{\xi} = 0$

Having the Table 2. 2-1 and the constraints (1), (2'), we can forget about the physical problem. All calculations which follow now are pure formalism.

a) LAGRANGE-equations of the 1st kind.

The formula is derived in the lecture :

$$P_j + \sum_k \lambda_k \varphi_{kj} = m_j \ddot{x}_j$$

φ_{kj} is identical with $\partial \varphi_k / \partial x_j$ in this case. Using Table 2. 2-1 we find out

$$(3) \qquad c(y - x) - \lambda_1 + 0 = m\ddot{x} ,$$

$$(4) \qquad c(x - y) + 0 + 0 = M\ddot{y} ,$$

$$(5) \qquad \sqrt{2}/2\, P + \lambda_1 + 0 = m\ddot{z} ,$$

$$(6) \qquad \sqrt{2}/2\, P + 0 + \lambda_2 = m\ddot{\xi} .$$

Equation (5) yields, when putting in $\ddot{z} = \ddot{x}$ (following from eq. (1)),

$$\lambda_1 = m\ddot{x} - \sqrt{2}/2\, P .$$

This can be inserted into eq. (3) :

$$c(y - x) - m\ddot{x} + \sqrt{2}/2\, P = m\ddot{x} .$$

This is a first equation of motion containing x and y only. The second equation of motion is eq. (4). Equation (6) is the only relation which contains λ_2, which, therefore, can be calculated by eq. (6), but this equation is of no interest if we want to calculate the equations of motion.

Rearranging the result we get

$$(7) \qquad \begin{bmatrix} 2m & 0 \\ 0 & M \end{bmatrix} \cdot \begin{bmatrix} \ddot{x} \\ \ddot{y} \end{bmatrix} + \begin{bmatrix} c & -c \\ -c & c \end{bmatrix} \cdot \begin{bmatrix} x \\ y \end{bmatrix} = \begin{bmatrix} \sqrt{2}/2\, P \\ 0 \end{bmatrix}$$

b) D'ALEMBERT's principle

Formula :

$$\delta' W = \sum_i \left(P_i - m_i \ddot{x}_i \right) \delta' x_i = 0$$

The relations between the $\delta' x_i$ can be found when differentiating φ_k with respect to the coordinates. From Table 2.2-1 we get P_i, m_i, x_i, so we can state :

$$\left[c(y-x) - m\ddot{x} \right] \delta' x + \left[c(x-y) - M\ddot{y} \right] \delta' y +$$

$$+ \left[\sqrt{2}/2\, P - m\ddot{z} \right] \delta' z + \left[\sqrt{2}/2\, P - 0 \right] \delta' \xi = 0$$

$\ddot{z}$ can be replaced by $\ddot{x}$ (constraint (1)); $\delta' x$ and $\delta' y$ are arbitrary values, whereas $\delta' z$ and $\delta' \xi$ are functions of $\delta' x$ and $\delta' y$: $\delta' z = \delta' x$; $\delta' \xi = 0$.

So we deduce :

$$\left[c(y-x) + \sqrt{2}/2\, P - 2m\ddot{x} \right] \delta' x + \left[c(x-y) - M\ddot{y} \right] \delta' y = 0 .$$

The coefficients of $\delta' x$ and $\delta' y$ must vanish, because these values are arbitrary. Thus, we find result (7).

c) GAUSS-principle.

c1) used as a variational principle

Formula :

$$\delta'' w = \sum_i \left(P_i - m_i \ddot{x}_i \right) \delta'' \ddot{x}_i = 0$$

In this form, the GAUSS-principle is identical with

d'ALEMBERT's principle, because the relations between the $\delta' x_i$ and the $\delta'' \ddot{x}_i$ are completely the same (cf. lecture course and Table 2.2-1).

c2) used as a minimum principle

$$\Gamma = \frac{1}{2}\left\{\left[\frac{c}{m}(y-x)-\ddot{x}\right]^2 m + \left[\frac{c}{M}(x-y)-\ddot{y}\right]^2 M + \right.$$

$$\left. + \left[\frac{\sqrt{2}}{2}\,\frac{P}{m} - \underbrace{\ddot{x}}_{\text{instead } \ddot{z}}\right]^2 m + \left[\frac{\sqrt{2}}{2}\,\frac{P}{m} - \underbrace{0}_{\text{instead } \ddot{\xi}}\right]^2 m\right\} \Longrightarrow \underbrace{\min}_{\text{GAUSS-Variation}}$$

In this formula the constraints (1), (2) are already used for replacing $z, \ddot{z}, \ddot{\xi}$ by $x, y, \ddot{x}, \ddot{y}$. Now we could use $\Gamma \Rightarrow \min$ for the numerical computation of $\ddot{x}$ and $\ddot{y}$ by an approximation method. But we must be careful. If the force P depends on $\ddot{x}, \ddot{y}$ in any way, it is not allowed to vary these values $\ddot{x}, \ddot{y}$, too.

Our problem is quite simple; therefore, nobody will calculate $\ddot{x}$ and $\ddot{y}$ numerically and approximately. The equations of motion can be found simply by partial derivations of Γ with respect to $\ddot{x}$ and $\ddot{y}$:

$$\frac{\partial \Gamma}{\partial \ddot{x}} = 0 \Longrightarrow \left[\ddot{x} - \frac{c(y-x)}{m}\right] m + \left[\ddot{x} - \frac{\sqrt{2}}{2}\,\frac{P}{m}\right] m = 0,$$

$$\frac{\partial \Gamma}{\partial \ddot{y}} = 0 \Longrightarrow \left[\ddot{y} - \frac{c}{M}(x-y)\right] M = 0 .$$

This is identical with result (7).

c3) The wrong principle (only for comparison)

In order to show the possible mistake we must introduce different masses instead of the two equal masses m. We assume that the left mass m of Fig. 2.2-1 is replaced by m' being different from m. Then, the correct solution is

$$(m + m')\ddot{x} + cx - cy = \frac{\sqrt{2}}{2} P, \tag{8}$$

$$M\ddot{y} - cx + cy = 0. \tag{9}$$

But a (wrong) principle of the form

$$\Gamma^* = \sum_j (P_j - m_j \ddot{x}_j)^2 \Longrightarrow \min \qquad \text{(GAUSS-Variation)}$$

would yield

$$mm\ddot{x} + m'm'\ddot{x} - mcy + mcx = m' \frac{\sqrt{2}}{2} P \tag{10}$$

instead of eq. (8). Multiplying eq. (8) by m and subtracting it from eq. (10) we get

$$(m' - m)\left(m'\ddot{x} - \frac{\sqrt{2}}{2} P\right) = 0,$$

which, for $m \neq m'$ leads to the wrong result : $\underline{m'\ddot{x} = \frac{\sqrt{2}}{2} P}$.

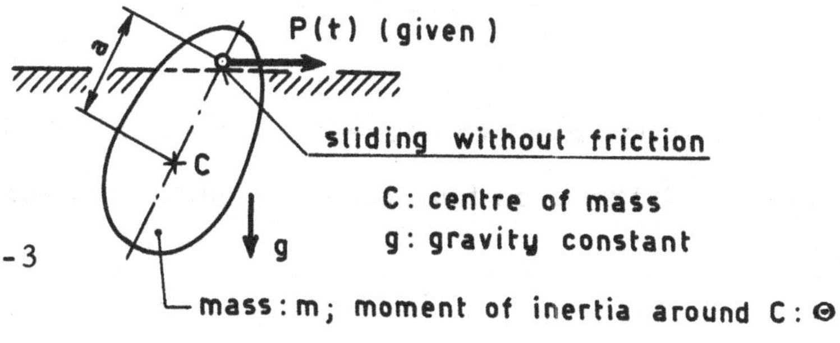

Fig. 2.2-3

Problem 2.2-2 : Find out the equations of motion for the system sketched in Fig. 2.2-3, by means of LAGRANGE equations of the 1st kind.

This problem has to demonstrate at first the introduction of a zero mass, at second the application of these methods to rotational motions.

At first, we draw a figure like Fig. 2.2-2 in order to find out the coordinates which have to be introduced.

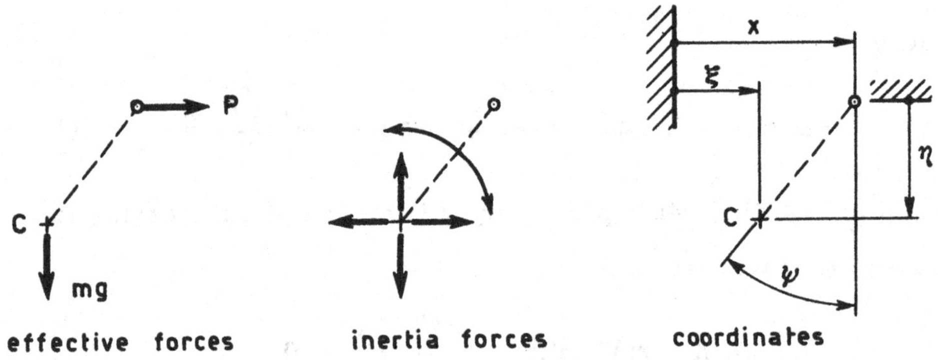

Fig. 2.2-4

The system has two degrees of freedom but, in Fig. 2.2-4, we have introduced four coordinates. So we need two constraints :

$$\left.\begin{aligned}\varphi_1 &\equiv \eta - a\cos\psi = 0 ;\\ \varphi_2 &\equiv \xi - x + a\sin\psi = 0 .\end{aligned}\right\} \tag{1}$$

The coordinate x describes a point, where no mass is situated. Therefore we have to add there a zero mass. Now we can print a table (Table 2.2-2).

Table 2.2-2

m_j	x_j	P_j	φ_{1j}	φ_{2j}
Θ	ψ	0	$a\sin\psi$	$a\cos\psi$
m	ξ	0	0	1
m	η	mg	1	0
0	x	P	0	-1

We use ψ and ξ as free coordinates. We start from the formula

$$P_j + \sum_k \lambda_k \varphi_{kj} = m_j \ddot{x}_j$$

and from the constraints (1). Then, we get

$$0 + \lambda_1 a\sin\psi + \lambda_2 a\cos\psi = \Theta\ddot{\psi} , \tag{2}$$

$$0 + 0 + \lambda_2 = m\ddot{\xi} , \tag{3}$$

(4) $$mg + \lambda_1 + 0 = m\ddot{\eta} ,$$

(5) $$P + 0 - \lambda_2 = 0 .$$

Equation (5) yields $\lambda_2 = P$. Putting this into eq. (3) we get

$$\underline{\underline{m\ddot{\xi} = P}}$$

as our first equation of motion. Substituting $\ddot{\eta}$ in eq. (4) by means of eqs. (1), eq. (4) yields

$$\lambda_1 = - m(g + a\dot{\psi}^2 \cos\psi + a\ddot{\psi}\sin\psi).$$

Inserting this and $\lambda_2 = P$ into eq. (2) we reach the second equation of motion :

$$\underline{\underline{(\Theta + ma^2 \sin^2\psi)\ddot{\psi} + m(a\dot{\psi}^2\cos\psi + g)a\sin\psi = Pa\cos\psi .}}$$

2.3. Problems with inholonomic constraints (cf. sect. 2. 2 of the lecture--notes).

The typical problems with inholomonic constraints are connected with wheels. So we will have them in our problem, too.

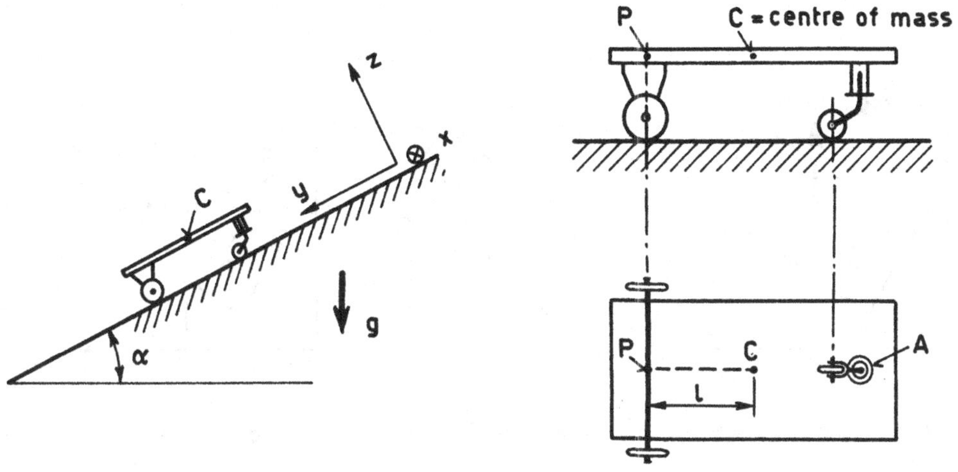

Fig. 2. 3-1

Problem 2. 3-1 : Find out equations of motion for the car sketched in Fig. 2. 3-1. The support at A can turn freely, so it does not restrict the motion of the car in the x, y -plane.

We start in the same way as we did in the last sub-section. Fig. 2. 3-2 shows the directions of the effective and the possible inertia forces as well as the coordinates entering our equations and several coordinates which will be used during the deduction of the constraint (see Fig. 2. 3-2-next page).

We use x, y and ψ in the equations. These coordinates are free, but the direction of the velocity $\underline{v}^P$ of point

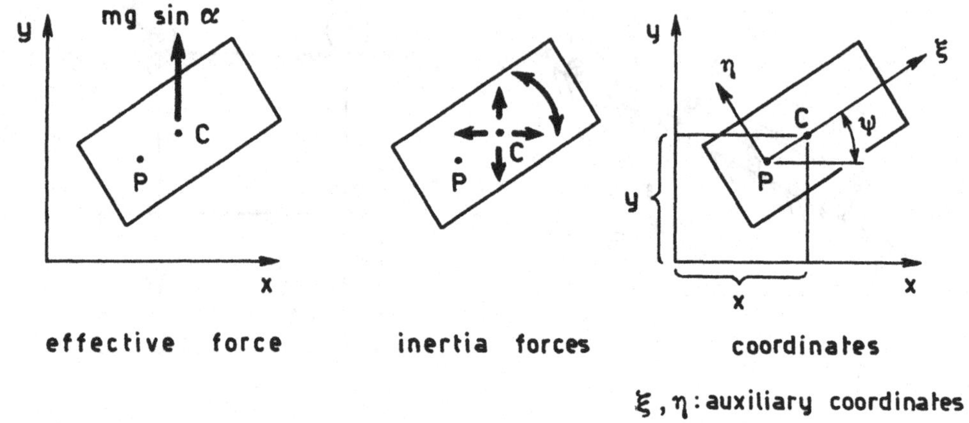

Fig. 2. 3-2

P is restricted (underlining denotes physical vectors) : $\underline{v}^P$ is parallel to the ξ-axis. Thus, we have to remark $v^P_\eta = 0$.

On the other hand, we can find out

$$\underline{v}^P = \underline{v}^C + \underline{\omega} \times \begin{bmatrix} -\ell \\ 0 \\ 0 \end{bmatrix}_{\xi,\eta,\zeta\text{ - system }(\zeta = z)}$$

where

$$\underline{v}^C = \begin{bmatrix} \dot{x} \\ \dot{y} \\ 0 \end{bmatrix}_{x,y,z} = \begin{bmatrix} \dot{x}\cos\psi + \dot{y}\sin\psi \\ -\dot{x}\sin\psi + \dot{y}\cos\psi \\ 0 \end{bmatrix}_{\xi,\eta,\zeta} ,$$

$$\underline{\omega} = \begin{bmatrix} 0 \\ 0 \\ \dot{\psi} \end{bmatrix}_{x,y,z} = \begin{bmatrix} 0 \\ 0 \\ \dot{\psi} \end{bmatrix}_{\xi,\eta,\zeta} .$$

Hence, we get

$$\underline{v}^P = \begin{bmatrix} \dot{x}\cos\psi + \dot{y}\sin\psi \\ -\dot{x}\sin\psi + \dot{y}\cos\psi \\ 0 \end{bmatrix}_{\xi,\eta,\zeta} + \begin{bmatrix} 0 \\ -\dot{\psi}\ell \\ 0 \end{bmatrix}_{\xi,\eta,\zeta} = \begin{bmatrix} v_\xi^P \\ 0 \\ 0 \end{bmatrix} .$$

Thus, we have the constraint :

$$\underline{-\dot{x}\sin\psi + y\cos\psi - \dot{\psi}\ell = v_\eta^P = 0 .} \tag{1}$$

This inholonomic constraint allowes the calculation of the quantities φ_{kj} of the LAGRANGE eqs. of the 1st kind. We want to use d'ALEMBERT's principle. For this, the $\delta' x_j$ have to satisfy the same relation as the velocities have to fulfil :

$$\delta'\psi = \frac{1}{\ell}(\delta' y \cos\psi - \delta' x \sin\psi) . \tag{2}$$

The usual table is printed in Table 2. 3-1.

Table 2. 3-1

m_j	x_j	P_j	φ_j
m	x	0	$\sin\psi$
m	y	$mg\cos\alpha$	$-\cos\psi$
Θ	ψ	0	ℓ

D'ALEMBERT's principle $\sum_i (m_i \ddot{x}_i - P_i)\delta' x_i = 0$

yields now

$$m\ddot{x}\,\delta'x + (m\ddot{y} - mg\sin\alpha)\,\delta'y + \Theta\ddot{\psi}\,\delta'\psi = 0$$

or with eq. (2)

$$\left(m\ddot{x} - \frac{\Theta}{l}\ddot{\psi}\sin\psi\right)\delta'x + \left(m\ddot{y} + \frac{\Theta}{l}\ddot{\psi}\cos\psi - mg\sin\alpha\right)\delta'y = 0,$$

where $\delta'x$ and $\delta'y$ are arbitrary. So we get

(3) $$m\ddot{x} - \frac{\Theta}{l}\ddot{\psi}\sin\psi = 0,$$

(4) $$m\ddot{y} - mg\sin\alpha + \frac{\Theta}{l}\ddot{\psi}\cos\psi = 0.$$

Equations (1), (3), and (4) are the complete set of equations of motion. By means of eq. (1), $\ddot{x}$ can be eliminated so that we get only two relations for y and ψ; the eliminated coordinate x can be calculated afterwards by eq. (1). So the final result is :

$$m\ddot{y} + \frac{\Theta}{l}\ddot{\psi}\cos\psi = mg\sin\alpha,$$

$$\left(\frac{\Theta}{l}\sin^3\psi + ml\sin\psi\right)\ddot{\psi} -$$

$$- m\sin\psi\cos\psi\,\ddot{y} = m\left(l\dot{\psi}^2\cos\psi - \dot{\psi}\dot{y}\right).$$

3. POTENTIALS (cf. sect. 2. 3 of the lecture-notes).

The following very simple example has to demonstrate that force-fields are possible(in an at least two-dimensional space)which are not potential force-fields, but which become potential fields if a restriction is satisfied.

Problem 3. -1 : A force $\underline{F}$ in a cartesian x, y -plane is given by $F_x = ax + by$, $F_y = cx + ey$. Check whether $\underline{F}$ is a potential force and calculate, if possible, the potential function.

If $\underline{F}$ is a potential force the relations

$$F_x = -\frac{\partial U}{\partial x}, \quad F_y = -\frac{\partial U}{\partial y}$$

have to be satisfied. Therefore, the integrability condition

$$\frac{\partial F_x}{\partial y} = -\frac{\partial^2 U}{\partial x \partial y} = -\frac{\partial^2 U}{\partial y \partial x} = \frac{\partial F_y}{\partial x}$$

must be fulfilled. This yields

$$\underline{\underline{b = c}}.$$

$\underline{F}$ is a potential force under this condition only.

Let us calculate the work done on a way as it is sketched in Fig. 3. -1. b and c may be arbitrary.

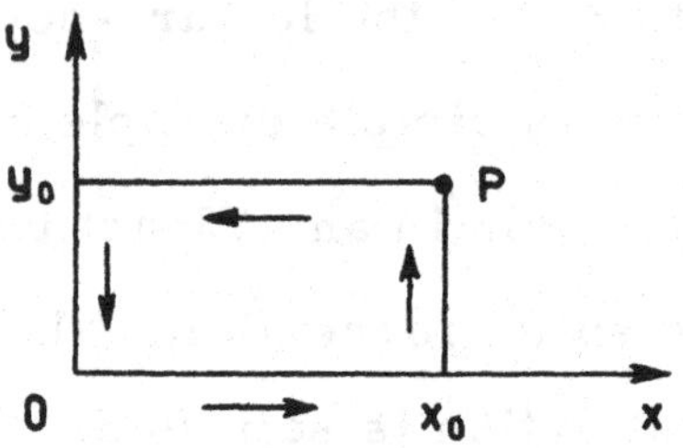

Fig. 3. -1

The total work done on the sketched way is the integral

$$\int F_x \, dx + \int F_y \, dy .$$

Thus, we get

$$W = \underbrace{\int_0^{x_0} a x \, dx + \int_0^{y_0} (c x_0 + e y) \, dy}_{\text{U at point P if a potential can be defined}} + \int_{x_0}^{0} (a x + b y_0) \, dx + \int_{y_0}^{0} e y \, dy =$$

$$= \overbrace{a \frac{x_0^2}{2} + c x_0 y_0 + e \frac{y_0^2}{2}} - a \frac{x_0^2}{2} - b x_0 y_0 - \frac{e y_0^2}{2} = (c - b) x_0 y_0 .$$

If $\underline{F}$ would be a potential force, W would be zero. So we see by this example, too, that $c - b = 0$ is necessary for a potential.

For $b = c$, U is described by

$$U = \frac{a}{2} x^2 + cxy + \frac{e}{2} y^2$$

if $U = 0$ at $x = y = 0$. The restriction $b = c$ is not necessary if only a specific line $y(x)$ in the x,y-plane is taken into account. On this line we always get potentials because of the restriction. So this problem is a very simple example for the effect that a restriction makes a non-potential force to a potential force.

4. DESCRIBING QUANTITIES OF CONTINUA(cf. sect. 2.4 of the lecture).

4.1. Tensor notation

Tensors are denoted in different manners. Therefore, it seems to be necessary to give a short introduction into the denotation which will be used in the next sub-section. We will use the normal tensor notation for the description of COSSERAT- continua, whereas Professor LIPPMANN used unit base vectors. The difference will be pointed out at the end of sect. 4.2.

a) Base-vectors.

We restrict our considerations to EUCLIDEAN three-dimensional spaces. There we can shift every vector parallel to itself without any difficulty. Therefore, we need only one base-vector system

$$\underline{e}_a \quad (a = 1,2,3)$$

Any vector $\underline{v}$ can be decomposed into components parallel to these base-vectors so that

$$\underline{v} = v_a \, \underline{e}_a \qquad (\text{summation of } a = 1,2,3)$$

is valid. In future, the base-vector systems will be distinguished by their indices : $a \ldots h$ will denote the basic EUCLIDEAN system. In this system we define sums and products and so on :

$$\underline{v} + \underline{w} \overset{!}{=} (v_a + w_a)\underline{e}_a .$$

Any type of product will have to satisfy

$$\underline{v} \circledast \underline{w} = v_a w_b (\underline{e}_a \circledast \underline{e}_b)$$

Furtheron we state that two special products have to be possible : the inner product defined by

$$\underline{e}_a \cdot \underline{e}_b = \delta_{ab} \text{ with } \delta_{ab} = \begin{cases} 1 & \text{if } a = b \\ 0 & \text{if } a \neq b \end{cases} \quad \text{KRONECKER-Symbol}$$

this leads us to

$$\underline{v} \cdot \underline{w} = v_a w_a)$$

and the cross product, defined by

$$\underline{e}_a \times \underline{e}_b = e_{abc}\, \underline{e}_c , \quad \text{where } e_{abc} = \begin{cases} +1 & \text{if } a,b,c = 1,2,3 \text{ or } 2,3,1 \text{ or } 3,1,2 \\ -1 & \text{if } a,b,c = 3,2,1 \text{ or } 2,1,3 \text{ or } 1,3,2 \\ 0 & \text{if two indices are equal.} \end{cases}$$

We derive

$$\underline{v} \times \underline{w} = v_a w_b e_{abc}\, \underline{e}_c .$$

For many applications, it seems to be better not to use the $\underline{e}_a$-system but an $\underline{g}_i$-system. Whereas the base-vectors $\underline{e}_a$ are orthonormal unit vectors, the $\underline{g}_i$ are not orthogonal to each other, have different length and, possibly, different dimensions. It is advantageous to define a second base-vector system $\underline{g}^i$ which is combined with the $\underline{g}_i$-system by the relation

$$\underline{g}_i \cdot \underline{g}^j = \delta_i^j = \begin{cases} 1 & \text{if } i = j \\ 0 & \text{if } i \neq j \end{cases} \quad \text{(another form of the KRONECKER-Symbol)}$$

The base-vectors $\underline{g}^i$ and $\underline{g}_i$ are special but usual vectors in the $\underline{e}_a$ -system :

$$\underline{g}_i = g_{ia}\,\underline{e}_a , \quad \underline{g}^j = g^j_a\,\underline{e}_a .$$

$$\underline{g}_i \cdot \underline{g}^j = \delta_i^j , \quad \text{therefore, yields}$$

$$g_{ia}\, g^j_a = \delta_i^j .$$

The base-vectors now are called "covariant" ($\underline{g}_i$) and "contravariant" ($\underline{g}^i$). If we want to describe an arbitrary vector $\underline{v}$ in any system $\underline{g}^i$, $\underline{g}_i$, or $\underline{e}_a$, we write

$$\underline{v} = v^i \underline{g}_i = v_i \underline{g}^i \quad (\text{summation over } i,j\ldots),$$

but

$$\underline{v} = v_a\,\underline{e}_a \quad (\text{summation over } a,b,\ldots,h).$$

It is quite unusual to make such a difference between orthonormal unit base-vectors and general base-vectors. But, in the authors' opinion, this is quite advantageous, as far as we restrict our considerations to EUCLIDEAN spaces.

If a vector has to be decomposed uniquely into components parallel to $\underline{g}_i$, the linear equation system for v^i

$$v^i g_{ia} = v_a$$

must have a unique solution. Thus,

$$\det (g_{ia}) \neq 0$$

is necessary. In the same way we find

$$\det (g^i_a) = \det^{-1}(g_{ia}) \neq 0 .$$

Further on we deduce

$$\underline{e}_a = e_{ai}\, \underline{g}^i \quad \text{leads to} \quad \underline{e}_a \cdot \underline{g}_j = e_{aj} ,$$

$$\underline{e}_a = e_a{}^i\, \underline{g}_i \quad \text{yields} \quad \underline{e}_a \cdot \underline{g}^j = e_a{}^j .$$

On the other hand we see :

$$\underline{g}_i \cdot \underline{e}_a = g_{ib}\, \underline{e}_b \cdot \underline{e}_a = g_{ia} , \quad \text{which shows} \quad g_{ia} = e_{ai} ,$$

$$\underline{g}^j \cdot \underline{e}_a = g^j_b\, \underline{e}_b \cdot \underline{e}_a = g^j{}_b , \quad \text{hence} \quad g^j_a = e_a{}^j .$$

The components of the $\underline{g}_i$-vectors in the $\underline{g}^i$-system are g_{ij}. $\underline{g}_i = g_{ij}\, \underline{g}^j$ yields $\underline{g}_i \cdot \underline{g}_j = g_{ik}\, \underline{g}^k \cdot \underline{g}_j = g_{ij} = g_{ji}$.

Similarly, we define g^{ij} and get :

$$\underline{g}^i = g^{ij}\, \underline{g}_j \quad \text{yelds} \quad g^{ij} = \underline{g}^i \cdot \underline{g}^j = g^{ji}.$$

From this, we deduce

$$v^i \underline{g}_i \cdot \underline{g}_\ell = v_k \underline{g}^k \cdot \underline{g}_\ell \quad \text{or}$$

$$v^i g_{i\ell} = v_\ell \quad \text{and} \quad v_i g^{ij} = v^j .$$

The product $\underline{v}\cdot\underline{w}$ now reads

$$v^i g_{ia}\, w_j\, g^j{}_a = v^i w_i \,, \text{ so we see :}$$

$$\underline{v}\cdot\underline{w} = v^i w_i = v^i g_{ij}\, w^j = v_j w^j = v_i g^{ij} w_j .$$

b) The cross-product.

$\underline{v}\times\underline{w} = \underline{x}$ reads

$$x_c = e_{abc}\, v_a w_b \,,$$

and, because of $x_c = x^i g_{ic}$ or $x_c g^j{}_c = x^i g_{ic}\, g^j{}_c = x^j$,

$$x^k = g^k{}_c\, x_c = e_{abc}\, v_a w_b\, g^k{}_c = v_i w_j\, g^i{}_a g^j{}_b g^k{}_c\, e_{abc} = \epsilon^{ijk} v_i w_j$$

or $x^k = \epsilon^{ijk} v_i w_j$, where $\epsilon^{ijk} = g^i{}_a g^j{}_b g^k{}_c\, e_{abc}$.

Correspondingly, we derive

$$\epsilon_{ijk} = g_{ia}\, g_{jb}\, g_{kc}\, e_{abc} \quad \text{and} \quad x_k = \epsilon_{ijk}\, v^i w^j .$$

These values ϵ^{ijk} satisfy

$$\epsilon^{ijk} = g^i{}_a g^j{}_b g^k{}_c\, e_{abc} = -\, g^i{}_a g^j{}_b g^k{}_c\, e_{bac} =$$

$$= -\, g^i{}_b g^j{}_a g^k{}_c\, e_{abc} = -\, g^j{}_a g^i{}_b g^k{}_c\, e_{abc} = -\, \epsilon^{jik} .$$

This yields $\epsilon^{ijk} = 0$ if $i = j$. Similarly we derive $\epsilon^{ijk} = -\epsilon^{ikj}$ and $\epsilon^{ijk} = -\epsilon^{kji}$. So we see :

$$\epsilon^{ijk} = 0 \quad \text{if} \quad i = j\,, \ \text{or}\ j = k\,, \ \text{or}\ i = k\,,$$

and for the six not-vanishing values :

$$\epsilon^{123} = \epsilon^{231} = \epsilon^{312} = -\epsilon^{213} = -\epsilon^{132} = -\epsilon^{321}.$$

We will not derive here :

$$\epsilon_{123} = \det\,(g_{ia})\,, \quad \epsilon^{123} = \det\,(g^{i}_{\ a}) = \epsilon^{-1}_{123}\,,$$

$$\epsilon_{123}\,\epsilon_{123} = \det\,(g_{ij})\,, \quad \epsilon^{123}\,\epsilon^{123} = \det(g^{ij})\,.$$

c) Tensors.

We define Tensors by

$$\underline{\underline{T}} = T_{abc\ldots}\,\overbrace{\underline{e}_a \circ \underline{e}_b \circ \underline{e}_c \circ \ldots}^{n\ \text{factor}}$$

where the "tensorial product" „o" yields 3^n independent (!) components :

$$\underline{\underline{T}} = \underline{\underline{U}} \quad \text{states} \quad T_{abc\ldots} = U_{abc\ldots}.$$

If $\underline{\underline{T}}$ is independent of the coordinate system, $\underline{\underline{T}}$ is called a "tensor". The transformation laws can be deduced from this definition.

$$T_{abc\cdots}\,\underline{e}_a\circ\underline{e}_b\circ\underline{e}_c\circ\ldots = T^{ij}{}_k\ldots\underline{g}_i\circ\underline{g}_j\circ\underline{g}^k\circ\ldots =$$
$$= T^{ij}{}_k\ldots g_{ia}\,g_{jb}\,g^k{}_c\ldots\underline{e}_a\circ\underline{e}_b\circ\underline{e}_c =$$
$$= T^{ij\ell}\ldots g_{\ell k}\,\underline{g}_i\circ\underline{g}_j\circ\underline{g}^k$$

yields

$$T_{abc\cdots} = T^{ij}{}_k\ldots\, g_{ia}\,g_{jb}\,g^k{}_c\ldots, \quad \text{and}$$
$$T^{ij}{}_k\ldots = T_{abc\cdots}\, g^i{}_a\,g^j{}_b\,g_{kc}\ldots, \quad \text{and}$$
$$T^{ij\ell}\ldots g_{k\ell} = T^{ij}{}_k\ldots \quad \text{etc.}$$

Besides the systems $\underline{e}_a$, $\underline{g}_i$ and $\underline{g}^i$, we can possibly use systems $\underline{g}^K, \underline{g}_K$. Then we define

$$\underline{g}_K = g_{Ki}\,\underline{g}_i\ ;\quad \underline{g}^K = g^{Ki}\,\underline{g}_i\ ;\quad \underline{g}^K = g^K{}_i\,\underline{g}^i\ ;\quad \underline{g}_K = g_K{}^i\,\underline{g}_i\ .$$

We derive

$$g_{Ki} = \underline{g}_K\cdot\underline{g}_i = g_{iK} \quad \text{etc.}\ .$$

$$T^{ij}{}_{k\ldots}\,\underline{g}_i\circ\underline{g}_j\circ\underline{g}^k\circ\ldots = T^{IJ}{}_K\ldots\underline{g}_I\circ\underline{g}_J\circ\underline{g}^K\circ\ldots =$$
$$= T^{IJ}{}_K\ldots g_I{}^i\,g_J{}^j\,g_k{}^K\ldots\underline{g}_i\circ\underline{g}_j\circ\underline{g}^k\circ\ldots$$

yields

$$T^{ij}{}_k\ldots = T^{IJ}{}_K\ldots g_I{}^i\,g_J{}^j\,g_k{}^K\ldots$$

This is the usual law of transformations for tensor co ordinates.

It is possible to handle "mixed tensors"

$$\underline{\underline{T}} = T^{iJ}{}_k\ldots\underline{g}_i\circ\underline{g}_J\circ\underline{g}^k\ldots = T^{ij}{}_k\ldots\underline{g}_i\circ\underline{g}_j\circ\underline{g}^k\ldots$$

etc. In this sense, the quantities $g_{ia}, g_{ij}, g_i^{\ j}, e_a^{\ j}, g^{ik}, g^i_{\ k}$ etc. can be treated to be special components of the tensor $\underline{\underline{E}} = \delta_{ab}\, \underline{e}_a \circ \underline{e}_b$.

Also the values e_{abc}, ϵ^{ijk}, and ϵ_{ijk} belong to one tensor $\underline{\underline{\epsilon}}$:

$$\underline{\underline{\epsilon}} = \epsilon^{ijk}\underline{g}_i \circ \underline{g}_j \circ \underline{g}_k = \epsilon_{ijk}\, \underline{g}^i \circ \underline{g}^j \circ \underline{g}^k = e_{abc}\, \underline{e}_a \circ \underline{e}_b \circ \underline{e}_c ,$$

what can be easily checked.

d) Products of tensors.

We will not define a cross-product for tensors, but a mixed inner- and tensorial product. We define this product in the $\underline{e}_a$-system and postulate that the result has to be independent of the base-vector system. Therefore, the result is a tensor, too.

Definition :

$\underline{\underline{T}} \otimes \underline{\underline{U}}$ with list $\begin{pmatrix} i \leftrightarrow j \\ \vdots \end{pmatrix}$ means : The base-vectors corresponding to the ith index of $\underline{\underline{T}}$ and the jth index of $\underline{\underline{U}}$ form an inner product. The base-vectors and coefficients which do not appear in the list, form a tensorial product.

This definition is, somehow, difficult and will be explained by an example :

$$\underline{\underline{T}} \text{ and } \underline{\underline{U}} \text{ may be given as}$$

$$\underline{\underline{T}} = T_{ab}\, \underline{e}_a \circ \underline{e}_b = T^{i}{}_{j}\, \underline{g}_i \circ \underline{g}^j = \ldots$$

$$\underline{\underline{U}} = U_{ab}\, \underline{e}_a \circ \underline{e}_b = U^{i}{}_{j}\, \underline{g}_i \circ \underline{g}^j = \ldots .$$

Then, we get

$$\begin{aligned}
\underline{\underline{T}} \underset{(1\leftrightarrow 1)}{\otimes} \underline{\underline{U}} &= \left(T_{ab}\, \underline{e}_a \circ \underline{e}_b\right) \underset{(1\leftrightarrow 1)}{\otimes} \left(U_{cd}\, \underline{e}_c \circ \underline{e}_d\right) \\
&= T_{ab} U_{cd} \left(\underline{e}_a \cdot \underline{e}_c\right)\left(\underline{e}_b \circ \underline{e}_d\right) \\
&= T_{ab} U_{cd}\, \delta_{ac}\, \underline{e}_b \circ \underline{e}_d \\
&= T_{ab} U_{ad}\, \underline{e}_b \circ \underline{e}_d \\
&= \left(T^{i}{}_{j}\, \underline{g}_i \circ \underline{g}^j\right) \underset{(1\leftrightarrow 1)}{\otimes} \left(U^{kl}\, \underline{g}_k \circ \underline{g}_l\right) \\
&= T^{i}{}_{j} U^{kl} g_{ik}\, \underline{g}^j \circ \underline{g}_l \\
&= T^{i}{}_{j} U_{i}{}^{l}\, \underline{g}^j \circ \underline{g}_l = T_{kj} U^{kl}\, \underline{g}^j \circ \underline{g}_l .
\end{aligned}$$

Special products can be denoted by points :

$$\underline{\underline{T}} \cdot \underline{\underline{U}} \overset{!}{=} \underline{\underline{T}} \underset{(\text{last}\leftrightarrow 1)}{\otimes} \underline{\underline{U}} = T_{ab} U_{bd}\, \underline{e}_a \circ \underline{e}_d = T^{i}{}_{j} U^{j}{}_{l}\, \underline{g}_i \circ \underline{g}^l$$

$$\underline{\underline{T}} \cdot\cdot \underline{\underline{U}} \stackrel{!}{=} \underline{\underline{T}} \underset{\begin{pmatrix} \text{last} \leftrightarrow 1 \\ \text{for - last} \leftrightarrow 2 \end{pmatrix}}{\otimes} \underline{\underline{U}} = T_{ab} U_{ba} = T^{i}{}_{j} U^{j}{}_{i} =$$

$$= T^{ij} U_{ji} = \dots .$$

A well-known example is :

$$\underline{\underline{\sigma}} = \underline{\underline{\underline{\underline{E}}}} \cdot\cdot \underline{\underline{\varepsilon}} \quad : \quad \sigma_{ab} = E_{abcd} \, \varepsilon_{dc} \, .$$

The cross-product of vectors can now be written

$$\underline{v} \times \underline{w} = (\underline{\underline{\underline{\epsilon}}} \cdot \underline{w}) \cdot \underline{v} = - \underline{v} \cdot (\underline{w} \cdot \underline{\underline{\underline{\epsilon}}}) \, .$$

e) Covariant derivatives.

Some times, for the description of the points in space, curvilinear coordinates ξ^i are used which means : the position vector $\underline{r}$ is given by

$$\underline{r} = r_a \underline{e}_a = \underline{r} (\xi^1, \xi^2, \xi^3) \, .$$

Normally, local base-vectors $\underline{g}_i$ are, then, used for the notation of tensors which belong to the different points of the field (tensor-fields). These local base-vectors can be defined arbitrarily, but it is usual to take

$$\underline{g}_i = \frac{\partial \underline{r}}{\partial \xi^i} = \frac{\partial r_a}{\partial \xi^i} \underline{e}_a \, , \quad \underline{g}^j = [g_{ij}]^{-1} \underline{g}_i = \frac{\partial \xi^j}{\partial r_a} \underline{e}_a .$$

If a tensor $\underline{\underline{T}}$, for example $\underline{\underline{T}} = T_{ab}\, \underline{e}_a \circ \underline{e}_b$ is given, the transformation law yields

$$T^i_{\;j\ldots} = T_{ab\ldots}\, g^i_{\;a}\, g_{jb\ldots} = T_{ab\ldots} \frac{\partial \xi^i}{\partial r_a} \frac{\partial r_b}{\partial \xi^j} \ldots = T^I_{\;J\ldots} \frac{\partial \xi^i}{\partial \eta^I} \frac{\partial \eta^J}{\partial \xi^j} \ldots .$$

The derivation with respect to one coordinate ξ^l yields :

$$\frac{\partial \underline{\underline{T}}}{\partial \xi^l} = \frac{\partial}{\partial \xi^l} (T^i_{\;k}\, \underline{g}_i \circ \underline{g}^k) = \frac{\partial}{\partial \xi^l} (T^i_{\;k}\, g_{ia}\, g^k_{\;b}\, \underline{e}_a \circ \underline{e}_b) =$$

$$= \frac{\partial}{\partial \xi^l} (T^i_{\;k}\, g_{ia}\, g^k_{\;b})\, \underline{e}_a \circ \underline{e}_b =$$

$$= \left(\frac{\partial T^i_{\;k}}{\partial \xi^l} g_{ia}\, g^k_{\;b} + T^i_{\;k} \frac{\partial g_{ia}}{\partial \xi^l} g^k_{\;b} + T^i_{\;k}\, g_{ia} \frac{\partial g^k_{\;b}}{\partial \xi^l} \right) \underline{e}_a \circ \underline{e}_b =$$

$$= \frac{\partial T^i_{\;k}}{\partial \xi^l} \underline{g}_i \circ \underline{g}^k + T^i_{\;k} \frac{\partial \underline{g}_i}{\partial \xi^l} \circ \underline{g}^k + T^i_{\;k}\, \underline{g}_i \circ \frac{\partial \underline{g}^k}{\partial \xi^l}$$

On the other hand, $\dfrac{\partial \underline{\underline{T}}}{\partial \xi^l}$ is a tensor of the same order as $\underline{\underline{T}}$ itself. Therefore, the expression

$$\frac{\partial \underline{\underline{T}}}{\partial \xi^l} = T^i_{\;k|l}\, \underline{g}_i \circ \underline{g}^k$$

can be defined, where $T^i_{\;k|l}$ is the so-called "covariant-derivative" of the tensor coordinate $T^i_{\;k}$.

For the evaluation of this covariant derivative we need the derivatives of $\underline{g}_i$ and $\underline{g}^k$, which are vectors and can be expressed by

$$\frac{\partial \underline{g}_i}{\partial \xi^l} = \left\{ {m \atop l\ i} \right\} \underline{g}_m \; ; \quad \frac{\partial \underline{g}^k}{\partial \xi^l} = \left[{k \atop l\ n} \right] \underline{g}^n ,$$

where these symbols $\left\{ {m \atop l\ i} \right\}$ and $\left[{k \atop l\ n} \right]$ are functions of the space coordinates and describe the geometry of the curvilinear coordinate system ξ^i.

We simply see that

$$\frac{\partial \underline{g}_i}{\partial \xi^l} \cdot \underline{g}^k = \left\{ {m \atop l\ i} \right\} \underline{g}_m \cdot \underline{g}^k = \left\{ {k \atop l\ i} \right\}$$

and

$$\frac{\partial \underline{g}^k}{\partial \xi^l} \cdot \underline{g}_i = \left[{k \atop l\ n} \right] \underline{g}^n \cdot \underline{g}_i = \left[{k \atop l\ i} \right] .$$

But $(\partial \underline{g}_i / \partial \xi^l) \cdot \underline{g}^k$ and $(\partial \underline{g}^k / \partial \xi^l) \cdot \underline{g}_i$ are the two parts of $\partial (\underline{g}_i \cdot \underline{g}^k) / \partial \xi^l$:

$$\left\{ {k \atop l\ i} \right\} + \left[{k \atop l\ i} \right] = \frac{\partial}{\partial \xi^l} (\underline{g}_i \cdot \underline{g}^k) = \frac{\partial}{\partial \xi^l} (\delta_i^k) = 0 .$$

So we need only one of these symbols. We take the first symbol which is equal to the CRISTOPHEL-symbol of the 2nd kind. The CRISTOPHEL-symbols of the 1st kind are of lower interest for our considerations. We would have to deal with them in non-EUCLIDian spaces.

We have now to write

$$\frac{\partial \underline{g}^k}{\partial \xi^\ell} = - \left\{ \begin{matrix} k \\ \ell\, n \end{matrix} \right\} \underline{g}^n .$$

Starting from $\underline{g}_i = \partial \underline{r} / \partial \xi^i$, we easily see that

$$\left\{ \begin{matrix} k \\ \ell\, i \end{matrix} \right\} = \frac{\partial^2 r_a}{\partial \xi^i \partial \xi^\ell} \cdot \frac{\partial \xi^k}{\partial r_a} = \frac{\partial^2 r_a}{\partial \xi^\ell \partial \xi^i} \cdot \frac{\partial \xi^k}{\partial r_a} = \left\{ \begin{matrix} k \\ n\, i \end{matrix} \right\}.$$

The covariant derivatives get the following form :

$$\underline{T^i{}_{k|\ell} = \frac{\partial}{\partial \xi^\ell} \left(T^i{}_k \right) + T^m{}_k \left\{ \begin{matrix} i \\ \ell\, m \end{matrix} \right\} - T^i{}_n \left\{ \begin{matrix} n \\ \ell\, j \end{matrix} \right\}.}$$

Examples are :

$$g_{ij|k} = \epsilon_{ijk|\ell} = 0 \quad \text{etc.}$$

These derivatives vanish, because g_{ij} and ϵ_{ijk} are coordinates of the constant tensors $\underline{\underline{E}}$ and $\underline{\underline{\underline{\epsilon}}}$ which do not depend on the space-coordinates. We will not prove here that the product law of differentiation holds for covariant derivatives but we will have to use it.

Example :

$$(v^{ik}{}_{\ell m}\, w_{jk}{}^{\ell}{}_{n}\, g^{mn})|_p =$$

$$= v^{ik}{}_{\ell m|p}\, w_{jk}{}^{\ell}{}_{n}\, g^{mn} + v^{ik}{}_{\ell m}\, w_{jk}{}^{\ell}{}_{n|p}\, g^{mn} + 0$$

We will not make use of the fact that $T^{i}{}_{k|\ell}\, \underline{g}_i \circ \underline{g}^k \circ \underline{g}^\ell$ is a tensor, too.

f) The GAUSS-GREEN-theorem.

When using the tensor notation explained above, the GAUSS-GREEN-theorem which was already mentioned in the lecture course, gets the form

$$\oint v^i \nu_i \, dB = \int^{I} v^i{}_{|i} \, dI ,$$

where $\underline{v} = v^i \underline{g}_i$ is any vector, and $\underline{\nu}$ is the normal unit vector on B directed to the outside of I .

4.2. The Cosserat-continuum

We want to apply the virtual work theorem to the theory of COSSERAT-continua in order to find out the describing quantities of this type of continua.

1. Starting point.

a) The principle

We want to apply the principle

$$\delta W_{in} = \delta W_{ex} + \delta W_{J}$$

or

$$\int^{I} Q < \delta' q > dI =$$

$$= \int^{I} f < \delta' x > dI + \oint^{B} S < \delta' x > dB - \int^{I} \ddot{x} m < \delta' x > dI$$

where Q_j are the internal forces, $\delta' q_j$ the internal virtual displacements, f_j external volume forces, S_j surface tractions, $\delta' x_j$ external virtual displacements, x_j the "true" displacements, and m a mass-matrix.

b) The COSSERAT-continuum

b1) In a COSSERAT-continuum we have to deal with a field of displacements (velocities v_j) and with rotational displacements (angular velocities ω_j) of the particles themselves. These two fields are independent from each other. So $\dot{x}$ has to be replaced by $\underline{v}$ and $\underline{\omega}$.

b2) At any surface we can find the usual stresses σ^{ij} and couple-stresses μ^{ij} : At any point of any surface we have to deal with a force per unit area which can be denoted by $\underline{\sigma}_{\overset{1}{a}}$ and a couple-stress $\underline{\mu}_{\overset{1}{a}}$ where these forces and couples are given in the local base-vector system $\underline{e}_a$ which has an $\underline{e}_{\overset{1}{a}}$ - unit-base-vector normal to the surface. $\underline{\sigma}_{\overset{1}{a}}$ and $\underline{\mu}_{\overset{1}{a}}$ can be expressed as $\underline{\sigma}_{\overset{1}{a}} = \sigma_{\overset{1}{a}b}\, \underline{e}_b$ and $\underline{\mu}_{\overset{1}{a}} = \mu_{\overset{1}{a}b}\, \underline{e}_b$, where σ_{ab} and μ_{ab} have to be tensor-coordinates. This is - at this place - a definition (cf. Fig. 4.2-1).

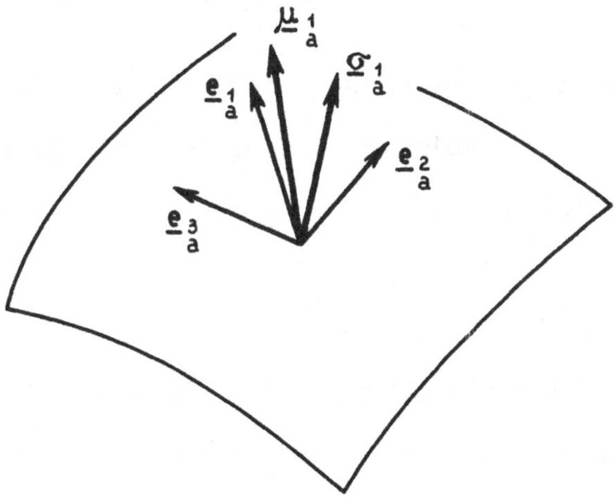

Fig. 4.2-1

This is valid at any arbitrary surface, especially at cross-sections inside the considered body, too. Therefore, the σ^{ij} and μ^{ij} are the internal forces, whereas the surface tractions are $\underline{\sigma}_{\substack{1\\a}}$ and $\underline{\mu}_{\substack{1\\a}}$. The normal unit vector $\underline{\nu}$ on B is $\underline{e}_{\substack{1\\a}}$.
Inside the body we can have volume forces $f^i \underline{g}_i = \underline{f}$ and torques per unit volume $m^j \underline{g}_j = \underline{m}$. They represent f in the principle.

b3) In a COSSERAT-continuum, $\delta W_{\mathfrak{J}}$ consists of two parts : The usual part $\int^{I} \varrho \dot{\underline{v}} \cdot \delta \underline{x}^{(trans)} \, dI$, where $\dot{\underline{v}}$ is the material acceleration $\underline{a}$. Correspondingly, we can introduce a rotational part :

$$\int^{I} (\underline{\underline{\vartheta}} \cdot \underline{\alpha}) \cdot \delta \underline{x}^{(rot)} \, dI ,$$

where $\underline{\underline{\vartheta}}$ is the tensor of inertia for the single particles per unit volume, and $\underline{\alpha}$ is the angular acceleration of the particles.

c) The variation

For our considerations, the δx are entirely free. Thus, we can use δx_j to be equal to $v_j \delta t$ and $\omega_j \delta t$, where v_j and ω_j are possible, but not necessarily true velocities or angular velocities, resp. δt is a time-variation. The velocities v_j and ω_j are arbitrary and - especially - independent from the internal and external forces.

Under these conditions, we define the internal displacement velocities which we call 'strain rates' and denote by λ_{ij} and $\varkappa_{ij}$ by the equation

$$\int^{I} Q \langle \delta q \rangle \, dI \overset{!}{=} \int^{I} (\underline{\underline{\sigma}} \cdot\cdot \underline{\underline{\lambda}} + \underline{\underline{\mu}} \cdot\cdot \underline{\underline{\varkappa}}) \, \delta t \, dI .$$

Now we can specialize the principle.

2. Specialized principle.

The specialized virtual work theorem reads

$$\int^{I}\left[\underline{\underline{\sigma}}\cdot\cdot\underline{\underline{\lambda}}+\underline{\underline{\mu}}\cdot\cdot\underline{\underline{\varkappa}}\right]\delta t\, dI =$$

$$=\int^{I}(\underline{f}\cdot\underline{v}+\underline{m}\cdot\underline{\omega})\,\delta t\, dI+\oint^{B}\left[\underline{\sigma}_{\underset{a}{1}}\cdot\underline{v}+\underline{\mu}_{\underset{a}{1}}\cdot\underline{\omega}\right]\delta t\, dB-$$

$$-\int^{I}\left[\varrho\,\underline{\underline{a}}\cdot\underline{v}+(\underline{\underline{\vartheta}}\cdot\underline{\alpha})\cdot\underline{\omega}\right]\delta t\, dI,$$

where $\underline{e}_{\underset{a}{1}}$ is equal to $\underline{\nu}$. The time variation δt appears at each term and is a constant. So it can be dropped. Using the tensor coordinates, we get :

$$\int^{I}(\sigma^{ij}\lambda_{ji}+\mu^{ij}\varkappa_{ji})\,dI=\int^{I}(f^{i}v_{i}+m^{i}\omega_{i})\,dI\;-$$

$$-\int^{I}(\varrho\, a^{i}v_{i}+\vartheta^{ij}\alpha_{j}\,\omega_{i})\,dI+\oint^{B}(\sigma_{\underset{a}{1}}{}^{i}v_{i}+\mu_{\underset{a}{1}}{}^{i}\omega_{i})\,dB.$$

It is necessary to replace the boundary integral. This is possible when we use

$$\sigma_{\underset{a}{1}}{}^{j}=e_{\underset{a}{1}k}\,\sigma^{kj}=\nu_{k}\sigma^{kj}\;;$$

$$\mu_{\underset{a}{1}}{}^{j}=\nu_{k}\,\mu^{kj}:$$

$$\oint^{B}(\sigma_{\underset{a}{1}}{}^{i}v_{i}+\mu_{\underset{a}{1}}{}^{i}\omega_{i})\,dB=\oint^{B}\nu_{j}(\sigma^{ji}v_{i}+\mu^{ji}\omega_{i})\,dB\,.$$

$(\sigma^{ji}v_{i}+\mu^{ji}\omega_{i})\,\underline{g}_{j}$ is a vector, therefore, we are we are allowed to apply the GAUSS-GREEN-theorem :

$$\oint^{B} \nu_j (\dots)\, dB = \int^{I} (\dots)_{|j}\, dI \,.$$

So we find :

$$\int^{I} \Big\{ (\sigma^{ij}\lambda_{ji} + \mu^{ij}\varkappa_{ji}) - (f^i v_i + m^i \omega_i) +$$

$$+ (\rho a^i v_i + \vartheta^{ij}\alpha_j \omega_i) - (\sigma^{ji} v_i + \mu^{ji}\omega_i)_{|j} \Big\} dI = 0$$

This integral has to vanish for every body or part of any body. Therefore, the integrand itself has to vanish. Thus, we get the basic equation :

$$\sigma^{ij}\lambda_{ji} + \mu^{ij}\varkappa_{ji} = f^i v_i + m^i \omega_i - \rho a^i v_i - \vartheta^{ij}\alpha_j \omega_i +$$

(1) $$+ \sigma^{ji}{}_{|j}\, v_i + \sigma^{ji} v_{i|j} + \mu^{ji}{}_{|j}\, \omega_i + \mu^{ji}\omega_{i|j} \,.$$

We remember that the left side was derived from the internal work and that v_i and ω_i where arbitrary possible velocities independent from the stresses and couple-stresses.

3. Pure translation

We are free to choose $\underline{v}$ and $\underline{\omega}$ in such a way that a pure (rigid body) translation is described. That means :

$\underline{v}$ = const., $\underline{\omega} \equiv 0$, hence, $v_j \neq 0$ arbitrary; $v_{j|k}$ =

$= \omega_j = \omega_{j|k} = 0 .$

Pure translation does not bring out internal work. So we get :

$$0 = f^i v_i + 0 - \varrho a^i v_i - 0 + \sigma^{ji}|_j v_i + 0 + 0 + 0$$

or

$$v_i \left(f^i - \varrho a^i + \sigma^{ji}|_j \right) = 0 .$$

This yields because of the fact that the v_i are arbitrary :

$$\underline{\underline{f^i - \varrho a^i + \sigma^{ji}|_j = 0 .}} \tag{2}$$

This is our first result.

4. Pure rotation.

Similarly to the pure translation, we can deal with pure rotations. This case is described by

$$\underline{\omega} = \text{const.} \quad \text{and} \quad d\underline{v} = \underline{\omega} \times d\underline{r}$$

($d\underline{v}$, $d\underline{r}$: local differentials). $\underline{\omega}$ = const. (but arbitrary) yields

$$\omega_i : \text{ arbitrary} ; \; \omega_{i|j} = 0 .$$

The second equation leads to

$$\frac{\partial \underline{v}}{\partial \xi^i} = \underline{\omega} \times \frac{\partial \underline{r}}{\partial \xi^i} = \underline{\omega} \times \underline{g}_i .$$

v_i itself remains arbitrary. $\partial \underline{v} / \partial \xi^i = \underline{\omega} \times \underline{g}_i$ can be expressed in the following manner :

$$v_{k|i}\, \underline{g}^k = \epsilon_{k\ell m}\, \omega^\ell g^m{}_i\, \underline{g}^k \quad \text{or}$$

(3)
$$\underline{v_{k|i} = \epsilon_{k\ell i}\, \omega^\ell = \epsilon_{k\ell i}\, g^{\ell i} \omega_j .}$$

Inserting this $v_{k|i}$ into eq. (1) we get

$$0 = (f^i - \varrho a^i + \sigma^{ji}|_j)\, v_i + m^i \omega_i - \vartheta^{ij} \alpha_j \omega_i +$$

$$+ \mu^{ji}|_j\, \omega_i + \sigma^{ji} \epsilon_{i\ell j}\, g^{\ell k} \omega_k$$

or with eq. (2)

$$\omega_i \left(m^i - \vartheta^{ij} \alpha_j + \mu^{ji}|_j + \sigma^{jk} \epsilon_{jk\ell}\, g^{\ell i} \right) = 0 .$$

This leads to

(4)
$$\underline{\underline{m^i - \vartheta^{ij} \alpha_j + \mu^{ji}|_j + \sigma^{jk} \epsilon_{jk\ell}\, g^{\ell i} = 0}}$$

This is a second result. The six equations (2), (4) are the equations of equilibrium.

5. Independence of $\delta' q_j$ from Q_j

At last, we use the independence of $\delta' q_j$ from Q_j to define λ_{ij} and $\varkappa_{ij}$ Eq. (1) yields together with eqs. (2), (4)

$$\sigma^{ij}\lambda_{ji} + \mu^{ij}\varkappa_{ji} = \left(m^i - \vartheta^{ij}\alpha_j + \mu^{ji}|_j\right)\omega_i + \sigma^{ji}v_{i|j} + \mu^{ji}\omega_{i|j} =$$

$$= -\sigma^{jk}\epsilon_{jkl}\, g^{li}\omega_i + \sigma^{ji}v_{i|j} + \mu^{ji}\omega_{i|j}.$$

After changing the indices and rearranging, this can be written

$$\sigma^{ij}\left(\lambda_{ji} + \epsilon_{ijk}\, g^{kl}\omega_l - v_{j|i}\right) + \mu^{ij}\left(\varkappa_{ji} - \omega_{j|i}\right) = 0 .$$

This must be valid for arbitrary values of σ^{ij}, μ^{ij}. Hence, the brackets have to vanish. So we find :

$$\underline{\underline{\lambda_{ji} = v_{j|i} - \epsilon_{ijk}\, g^{kl}\omega_l}} , \tag{5}$$

$$\underline{\underline{\varkappa_{ji} = \omega_{j|i}}} . \tag{6}$$

These are the last results.

By our considerations, we have determined strain-rates. Now the question arises whether these strain rates are rates of strains. At classical continua, such a connection is possible. But a similar definition of strains in the case of COSSERAT-continua is impossible, because angular velocities can not be defined as rates of any angles. This is the most important difficulty of the theory of elasticity of polar continua.

6. Comparison of the results with equations for classical continua.

Equation (2) is exactly the equation of equilibrium for classical continua.

At classical continua, m^i, ϑ^{ij}, and μ^{ij} do not exist. Therefore, eq. (4) yields

$$\sigma^{jk} \epsilon_{jkl} g^{li} = 0$$

which is identical with

$$\sigma_{jk} \epsilon^{jki} = 0 .$$

This shows $\sigma_{jk} = \sigma_{kj}$ and, hence, $\sigma^{jk} = \sigma^{kj}$.

Now, we want to compare the compatibility condition (5) with classical compatibility relations. We must recognize that in the case of classical continua the rotation is combined with the velocity field. $\underline{\omega}$ is the average angular velocity with the coordinates

$$\omega_l = \frac{1}{2} g_{lm} \epsilon^{mnp} v_{p|n}$$

(this will not be derived here). Inserting this into eq. (5) we get

$$\lambda_{ji} = v_{j|i} - \frac{1}{2} \epsilon_{ijk} g^{kl} g_{lm} \epsilon^{mnp} v_{p|n} =$$

$$= v_{j|i} - \frac{1}{2} \epsilon_{kij} \epsilon^{knp} v_{p|n} =$$

$$= v_{j|i} - \frac{1}{2} (\delta_i^n \delta_j^p - \delta_i^p \delta_j^n) v_{p|n} =$$

$$= v_{j|i} - \frac{1}{2}\left(v_{j|i} - v_{i|j}\right),$$

$$\lambda_{ji} = \frac{1}{2}\left(v_{i|j} + v_{j|i}\right).$$

This is well-known.

The last compatibility condition (6) is of no sense in the case of classical continua, because neither $\varkappa_{ji}$ nor $\omega_{j|i}$ are declared.

7. Comparison with formulae derived in the lecture-course.

In this section we will compare the equations of equilibrium for cylindrical coordinates, derived in the lecture-course and following from eq. (2). We simplify this task by assuming that no volume forces appear. So we have to compare

$$\sigma^{ij}|_i = 0$$

and the equations of the lecture-course without internal volume forces and acceleration terms.

First, we have to define the coordinates $\xi^i = r, \vartheta, z$. Hence, $\underline{r}$ is given by

$$\underline{r} = r\cos\vartheta\, \underset{a}{\underline{e}_1} + r\sin\vartheta\, \underset{a}{\underline{e}_2} + z\, \underset{a}{\underline{e}_3}$$

which shall be identical with the notation

$$\underline{r} = \begin{bmatrix} r\cos\vartheta \\ r\sin\vartheta \\ z \end{bmatrix}.$$

This yields

$$\underline{g}_1 = \begin{bmatrix} \cos\vartheta \\ \sin\vartheta \\ 0 \end{bmatrix} \qquad \underline{g}_2 = \begin{bmatrix} -r\sin\vartheta \\ r\cos\vartheta \\ 0 \end{bmatrix} \qquad \underline{g}_3 = \begin{bmatrix} 0 \\ 0 \\ 1 \end{bmatrix}.$$

Thus, the matrix of the quantities g_{ij} can be given as

$$\left[g_{ij}\right] = \begin{bmatrix} 1 & 0 & 0 \\ 0 & r^2 & 0 \\ 0 & 0 & 1 \end{bmatrix}$$

which yields easily

$$\left[g^{ij}\right] = \left[g_{ij}\right]^{-1} = \begin{bmatrix} 1 & 0 & 0 \\ 0 & r^{-2} & 0 \\ 0 & 0 & 1 \end{bmatrix}$$

and

$$\underline{g}^1 = \underline{g}_1 \; ; \quad \underline{g}^3 = \underline{g}_3 \; ; \quad \underline{g}^2 = \begin{bmatrix} -\sin\vartheta/r \\ \cos\vartheta/r \\ 0 \end{bmatrix}.$$

Having these vectors $\underline{g}_i$, $\underline{g}^i$, we can compute the CRISTOPHEL-symbols.

$$\left\{ {2 \atop 1\,2} \right\} = \left\{ {2 \atop 2\,1} \right\} = \frac{1}{r} \; ; \quad \left\{ {1 \atop 2\,2} \right\} = -r .$$

The other CRISTOPHEL-symbols vanish.

The equation $\sigma^{ij}|_i = 0$ can be written as

$$\sigma^{ij}|_i = \frac{\partial \sigma^{ij}}{\partial \xi^i} + \begin{Bmatrix} i \\ i \ \ell \end{Bmatrix} \sigma^{\ell j} + \begin{Bmatrix} j \\ i \ \ell \end{Bmatrix} \sigma^{i\ell} = 0 . \tag{7}$$

For $j = 1$ $(j \hat{=} r)$, we get :

$$\frac{\partial \sigma^{rr}}{\partial r} + \frac{\partial \sigma^{\vartheta r}}{\partial \vartheta} + \frac{\partial \sigma^{zr}}{\partial z} + \frac{\sigma^{rr}}{r} - r\,\sigma^{\vartheta\vartheta} = 0, \tag{8}$$

whereas, in the lecture-course, the result was

$$\frac{\partial \bar{\sigma}_{rr}}{\partial r} + \frac{1}{r}\frac{\partial \bar{\sigma}_{\vartheta r}}{\partial \vartheta} + \frac{\partial \bar{\sigma}_{rz}}{\partial z} + \frac{\bar{\sigma}_{rr} - \bar{\sigma}_{\vartheta\vartheta}}{r} = 0 \tag{9}$$

($\bar{\sigma}_{ij}$ is used to distinguish the notations).

As we see, the equations are not identical. The reason is that Professor LIPPMANN dealt with unit base-vectors whereas our $\underline{g}_2$ is no unit vector. So we get the transformation rules

$$\left.\begin{array}{c}
\sigma^{rr} = \bar{\sigma}_{rr}\,, \quad \sigma^{r\vartheta} = \sigma^{\vartheta r} = \frac{1}{r}\bar{\sigma}_{r\vartheta} = \frac{1}{r}\bar{\sigma}_{\vartheta r}\,, \\
\sigma^{rz} = \sigma^{zr} = \bar{\sigma}_{rz} = \bar{\sigma}_{zr}\,, \\
\sigma^{\vartheta\vartheta} = \frac{1}{r^2}\bar{\sigma}_{\vartheta\vartheta}\,, \\
\sigma^{\vartheta z} = \sigma^{z\vartheta} = \frac{\bar{\sigma}_{\vartheta z}}{r} = \frac{1}{r}\bar{\sigma}_{z\vartheta}\,, \quad \sigma^{zz} = \bar{\sigma}_{zz}\,.
\end{array}\right\} \tag{10}$$

By means of this relation, eq. (8) yields exactly eq. (9). For $j = 2$ we get from eq. (7) :

$$\frac{\partial \sigma^{r\vartheta}}{\partial r} + \frac{\partial \sigma^{\vartheta\vartheta}}{\partial \vartheta} + \frac{\partial \sigma^{z\vartheta}}{\partial z} + 3\,\frac{\sigma^{r\vartheta}}{r} = 0, \tag{11}$$

whereas we know from the lecture

$$(12)\qquad \frac{\partial\bar{\sigma}_{r\vartheta}}{\partial r} + \frac{1}{r}\,\frac{\partial\bar{\sigma}_{\vartheta\vartheta}}{\partial\vartheta} + \frac{\partial\bar{\sigma}_{z\vartheta}}{\partial z} + 2\,\frac{\bar{\sigma}_{rz}}{r} = 0 .$$

The most interesting difference is that the coefficients at the last term differ. But, by the transformation rules (10), eq. (11) leads to eq. (12) : One term $-\bar{\sigma}_{r\vartheta}/r$ is derived from the first term. The result of the transformation is eq. (12) multiplied by r^{-1}.

At last, the equations for $j = 3$ have to be compared :

$$\frac{\partial\sigma^{rz}}{\partial r} + \frac{\partial\sigma^{\vartheta z}}{\partial\vartheta} + \frac{\partial\sigma^{zz}}{\partial z} + \frac{\sigma^{rz}}{r} = 0,$$

$$\frac{\partial\bar{\sigma}_{rz}}{\partial r} + \frac{1}{r}\,\frac{\partial\sigma_{\vartheta z}}{\partial\vartheta} - \frac{\partial\bar{\sigma}_{zz}}{\partial z} + \frac{\sigma_{rz}}{r} = 0 .$$

They are quite similar and become identical by means of eqs. (10).

5. PASSIVE WORK (cf. sect. 2.5 of the lecture).

Under this headline a problem of the theory of deflected beams will be solved by means of dummy loads and dummy displacements.

Problem 5.-1 : Find out the displacement $\hat{v}$ and the inclination angle $\hat{\omega}$ at the point where $\hat{P}$ and $\hat{M}$ are acting for the problem sketched in Fig. 5.-1 :

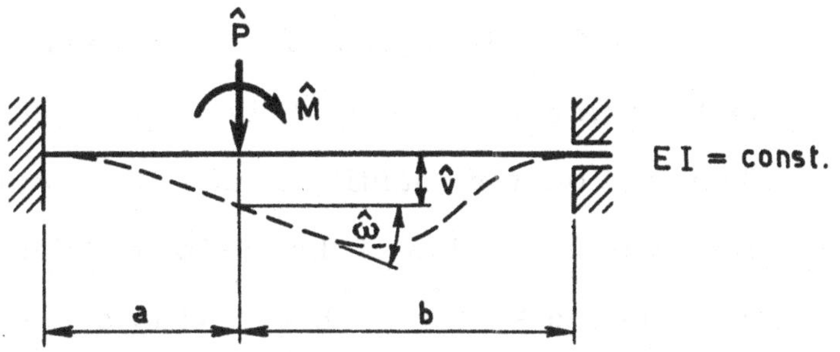

Fig. 5.-1

5.1. Dummy loads

The basic equation is

$$\Lambda_{ex}(F, y) = \Lambda_{in}(Q, r(\hat{Q})).$$

It is derived in the lecture. We specify it for the aplication to the theory of deflected beams :

$$\sum_i P_i \Delta\hat{x}_i + \sum_j M_j \hat{\psi}_j + \left(\int q_k \Delta\hat{x}_k \, d\xi\right) =$$

$$= \int \left\{ \frac{N\hat{N}}{EA} + \varkappa_i \frac{Q_i\hat{Q}_i}{GA} + \frac{M_T\hat{M}_T}{GI_T} + \frac{M_{bi}\hat{M}_{bi}}{EI_i} \right\} d\xi + \frac{P_c\hat{P}_c}{c} + \dots,$$

where P_i and M_j are single external loads, the corresponding displacements are Δx_i, ψ_j ; q_k is a load per unit length, N is the normal- Q_i the shear- (internal) force, M_T the torsion-, and M_{bi} a bending-moment.

EA, $\frac{1}{\varkappa_i} GA$, GI_T, and EI_i are tension-, shear-, torsion- and bending-stiffness. P_c is a spring force.

The roof on any quantity denotes the true state.

For the application of the basic equation, we need the true statics. But the system of Fig. 5.-1 is statically twice undeterminate. So we have to make it statically determinate by introducing reactions as external loads :

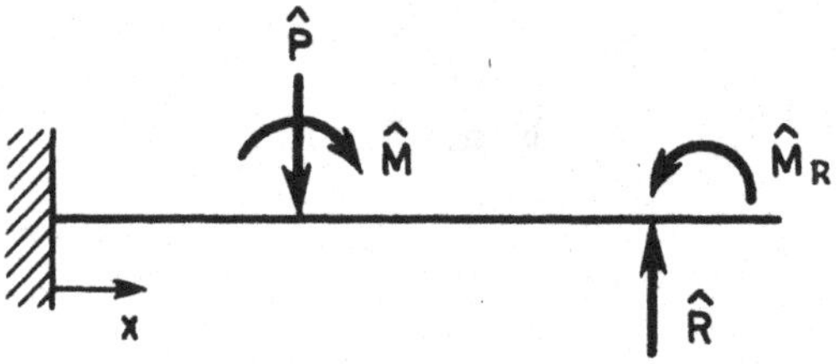

Fig. 5. -2

Fig.5.-2 shows the system which will be treated. If this has to be correct, we must postulate

$$\Delta \hat{x}_R = \hat{\psi}_{M_R} = 0.$$

Furthermore, we introduce a dummy load (P, M, R, M_R) parallel to each force or moment. We will restrict our considerations to pure bending, so we get the relation :

$$P\hat{v} + R\Delta\hat{x}_R + M\hat{\omega} + M_R\hat{\psi}_{MR} =$$

$$= \int_0^{a+b} \frac{1}{EI} M_b(P, R, M, M_R)\,\hat{M}_b(\hat{P}, \hat{R}, \hat{M}, \hat{M}_R)\,dx \qquad (1)$$

M_b and $\hat{M}_b$ are sums of the bending-moments which are printed in Fig. 5.-3 and which belong to cases where only one of the four loads is acting. The figure is printed for P, M, R, M_R being positive. The picture makes no difference between the dummy and the true loads.

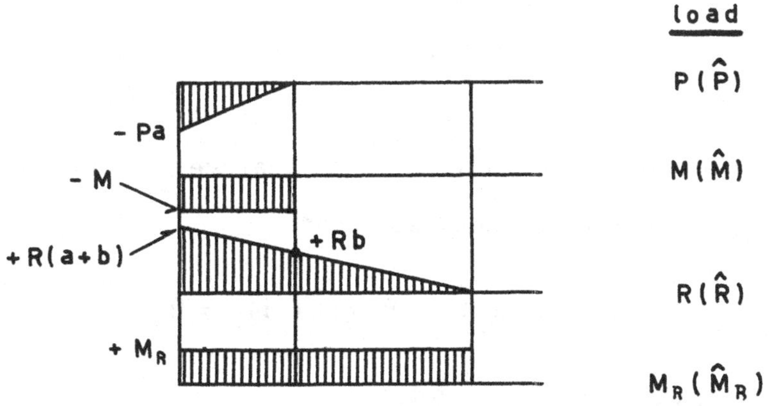

Fig. 5.-3

When applying the theory of dummy loads, we use advantageously the integral

$$(2) \qquad \frac{M_b \hat{M}_b}{EI} \, dx = \frac{\ell}{EI} \left\{ \frac{M_1 \hat{M}_1}{3} + \frac{1}{6} \left(M_1 \hat{M}_2 + M_2 \hat{M}_1 \right) + \frac{M_2 \hat{M}_2}{3} \right\},$$

where M_b and $\hat{M}_b$ have to be linear functions of x between $x = 0$ and $x = \ell$, and where $M_b(0) = M_1$, $\hat{M}_b(0) = \hat{M}_1$, $M_b(\ell) = \hat{M}_2$, $\hat{M}_b(\ell) = \hat{M}_2$ (cf. Fig. 5.-4).

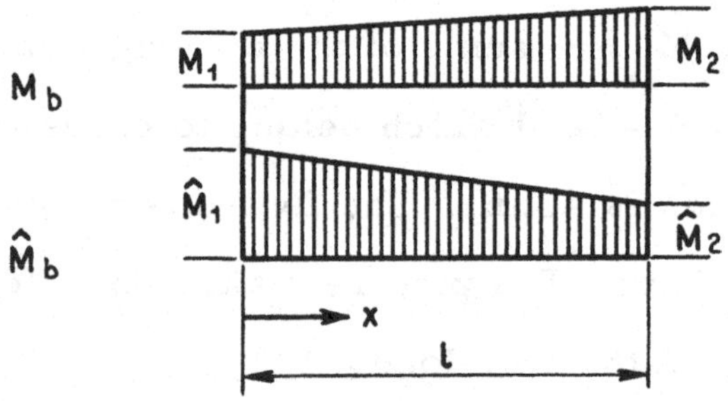

Fig. 5.-4

We use eq. (1) in such a way that we get four separate equations. We start with the case that

$$R \neq 0, \quad \text{but} \quad P = M = M_R = 0 .$$

Equation (1) then reads

$$R \, \Delta \hat{x}_R = \frac{1}{EI} \cdot \left\{ a \left[\frac{1}{3} R \, (a+b) \left(-\hat{P} a - \hat{M} + \hat{R} \, (a+b) + \hat{M}_R \right) + \right.\right.$$

$$+ \frac{1}{6} R (a+b) (-\hat{M} + \hat{R} b + \hat{M}_R) +$$

$$+ \frac{1}{6} R b (-\hat{P} a - \hat{M} + \hat{R}(a+b) + \hat{M}_R) + \frac{1}{3} R b (-\hat{M} + \hat{R} b + \hat{M}_R)\Big] +$$

$$+ b \left[\frac{1}{3} R b (\hat{R} b + \hat{M}_R) + \frac{1}{6} R b \hat{M}_R\right]\Big\}.$$

$\Delta \hat{x}_R$ has to be zero. R appears in each term, so it can be dropped. Then we get

$$\hat{R} \cdot \frac{1}{3} (a+b)^3 + \hat{M}_R \cdot \frac{1}{2} (a+b)^2 =$$

$$= \hat{P} \left(\frac{a^3}{3} + \frac{a^2 b}{2}\right) + \hat{M} \left(ab + \frac{a^2}{2}\right) . \tag{3}$$

Similarly, we derive from $M_R \hat{\psi}_{M_R} = 0$:

$$\hat{R} \cdot \frac{1}{2} (a+b)^2 + \hat{M}_R \cdot (a+b) = \hat{P} \frac{a^2}{2} + \hat{M} a . \tag{4}$$

Taking $P \neq 0$, $R = M_R = M = 0$, we find out

$$EI \hat{v} P = P(-a^2) \left[\frac{1}{3} (-\hat{P} a - \hat{M} + \hat{R}(a+b) + \hat{M}_R) + \right.$$

$$\left. + \frac{1}{6} (-\hat{M} + \hat{R} b + \hat{M}_R)\right]$$

or

$$EI \hat{v} = \frac{\hat{P} a^3}{3} + \frac{\hat{M} a^2}{2} - \hat{R} a^2 \left(\frac{a}{3} + \frac{b}{2}\right) - \frac{\hat{M}_R a^2}{2} . \tag{5}$$

Correspondingly, we get from $M \neq 0$; $P = R = M_R = 0$:

$$EI \hat{\omega} = + \frac{\hat{P} a^2}{2} + \hat{M} a - \hat{R} \left(\frac{a^2}{2} + ab\right) - \hat{M}_R a . \tag{6}$$

Eqs. (3) and (4) enable us to calculate $\hat{R}$ and $\hat{M}_R$ as functions of $\hat{M}$ and $\hat{P}$:

(7) $$\hat{M}_R = (a+b)^{-2} \cdot \left[-\hat{P} a^2 b + \hat{M}(a^2 - 2ab) \right],$$

(8) $$R = (a+b)^{-3} \cdot \left[\hat{P}(a^3 + 3a^2 b) + 6\hat{M} ab \right].$$

These expressions for $\hat{R}$ and $\hat{M}_R$ can be inserted into eqs. (5), (6). Then, we get the final result

(9) $$\hat{v} = \frac{a^2 b^2}{EI(a+b)^3} \left[\hat{P} \frac{ab}{3} + \hat{M} \frac{b-a}{2} \right],$$

(10) $$\hat{\omega} = \frac{ab}{EI(a+b)^3} \left[\hat{P} ab \frac{b-a}{2} + M(a^2 - ab + b^2) \right].$$

Possibly, the question could have been to calculate $\hat{P}$ and $\hat{M}$, whereas $\hat{v}$ and $\hat{\omega}$ would have been given. In this case eqs. (9), (10) had to be inverted. But we will see soon that for this problem it is easier to apply the method of dummy displacements.

5.2. Dummy displacements

The basic equation for the theory of dummy displacements is

$$\Lambda_{ex}(\hat{F}, y) = \Lambda_{in}(Q(\hat{r}), r)$$

or, specialized to the theory of beams :

$$\sum_i \hat{P}_i \,\Delta x_i + \sum_j \hat{M}_j \psi_j + \left(\int^{\ell} \hat{q}_k \,\Delta x_k \,d\xi\right) =$$

$$= \int^{\ell} \left\{ E A \hat{\varepsilon} \varepsilon + \ldots + E I_i \hat{w}_i'' w_i'' \right\} d\xi + c \,\Delta \ell \,\hat{\Delta \ell} + \ldots \qquad (11)$$

These relations are useful only in cases where the true values of $\hat{r}$ can be written down as unique functions of a relatively small number of parameters. For straight beams loaded by single forces and moments and with constant bending stiffness, $\hat{w}''$ is proportional to $\hat{M}_b$ which is linear. Thus, $\hat{w}$ itself is a cubic function of the beam coordinates. So we can express $\hat{w}''$ by means of $\hat{w}$ and $\hat{w}'$ at the ends of unloaded parts of the beams.

Let us apply this theory to problem 5. -1.

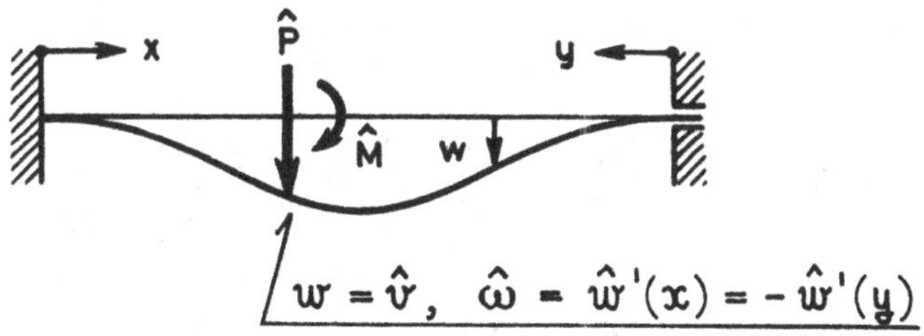

Fig. 5. -5

According to Fig. 5. -5 we introduce the coordinates x and y. The displacements are w. At that point, where $\hat{P}$

and $\hat{M}$ are acting, we have

$$(12) \qquad \hat{w} = \hat{v} \quad \text{and} \quad \hat{w}'(x) = \hat{\omega}\,, \quad \hat{w}'(y) = -\hat{\omega}\,.$$

w as a function of x or y resp., has to satisfy the boundary conditions $w(0) = w'(0) = 0$. This is given if we use

$$w(x) = Ax^2 + Bx^3\,; \qquad w(y) = Cy^2 + Dy^3.$$

The boundary conditions (12) prescribe the parameters A, B, C, D. So the functions $\hat{w}(x)$ and $\hat{w}(y)$ become

$$\hat{w}(x) = \left(\frac{3}{a^2}x^2 - \frac{2}{a^3}x^3\right)\hat{v} - \left(\frac{1}{a}x^2 - \frac{1}{a^2}x^3\right)\hat{\omega}\,,$$

$$\hat{w}(y) = \left(\frac{3}{b^2}y^2 - \frac{2}{b^3}y^3\right)\hat{v} + \left(\frac{1}{b}y^2 - \frac{1}{b^2}y^3\right)\hat{\omega}$$

or

$$(13) \quad \begin{cases} \hat{w}''(x) = \left(\dfrac{6}{a^2} - \dfrac{12}{a^3}x\right)\hat{v} - \left(\dfrac{2}{a} - \dfrac{6}{a^2}x\right)\hat{\omega}\,, \\[2ex] \hat{w}''(y) = \left(\dfrac{6}{b^2} - \dfrac{12}{b^3}y\right)\hat{v} + \left(\dfrac{2}{b} - \dfrac{6}{b^2}y\right)\hat{\omega}\,. \end{cases}$$

$w(x, v, \omega)$ and $w(y, v, \omega)$ are chosen like $\hat{w}$: eqs. (13) without roofs. This yields with eq. (11) :

$$\hat{M}\omega + \hat{P}v = EI \cdot \left\{ \int_0^a \hat{w}''(x)\, w''(x)\, dx + \int_0^b \hat{w}''(y)\, w''(y)\, dy \right\} =$$

$$= EI \cdot \left\{ \int_0^a \left[\left(\frac{36}{a^4} - \frac{144x}{a^5} + \frac{144}{a^6} x^2 \right) v\hat{v} + \left(\frac{4}{a^2} - \frac{24}{a^3} x + \frac{36}{a^4} x^2 \right) \omega\hat{\omega} - \right.\right.$$

$$\left. - \left(\frac{12}{a^3} - \frac{60}{a^4} x + \frac{72}{a^5} x^2 \right) \left(v\hat{\omega} + \omega\hat{v} \right) \right] dx +$$

$$+ \int_0^b \left[\left(\frac{36}{b^4} - \frac{144}{b^5} y + \frac{144}{b^6} y^2 \right) v\hat{v} + \left(\frac{4}{a^2} - \frac{24}{a^3} y + \frac{36}{a^4} y^2 \right) \omega\hat{\omega} + \right.$$

$$\left.\left. + \left(\frac{12}{b^3} - \frac{60}{b^4} y + \frac{72}{b^5} y^2 \right) \left(v\hat{\omega} + \omega\hat{v} \right) \right] dy \right\} =$$

$$= EI \cdot \left\{ 12 \left(\frac{1}{a^3} + \frac{1}{b^3} \right) v\hat{v} + 4 \left(\frac{1}{a} + \frac{1}{b} \right) \omega\hat{\omega} - \right.$$

$$\left. - 6 \left(\frac{1}{a^2} - \frac{1}{b^2} \right) \left(v\hat{\omega} + \omega\hat{v} \right) \right\}. \qquad (14)$$

ω and v are arbitrary. Therefore, the equation (14) is equiv alent to two equations :

$$v = 0, \quad \omega \neq 0 : \quad \hat{M} = EI \left\{ 4 \left(\frac{1}{a} + \frac{1}{b} \right) \hat{\omega} - 6 \left(\frac{1}{a^2} - \frac{1}{b^2} \right) \hat{v} \right\}$$

and :

$$\omega = 0\,,\ \upsilon \neq 0\ :\ \hat{P} = EI\left\{12\left(\frac{1}{a^3} + \frac{1}{b^3}\right)\hat{\upsilon} - 6\left(\frac{1}{a^2} - \frac{1}{b^2}\right)\hat{\omega}\right\}.$$

As we see, the use of dummy displacements is much easier than the use of dummy loads in this case.

At the end of this section some remarks are to be made on the methods of dummy loads and dummy displacements applied to other problems of beam theory.

1. It is possible to find out the internal forces at definite points of a beam in a statically undeterminate system by introducing links. The system with links has to be statically determinate. The work of the corresponding dummy load is zero. We see this easily from Fig. 5. -6 for a bending-moment.

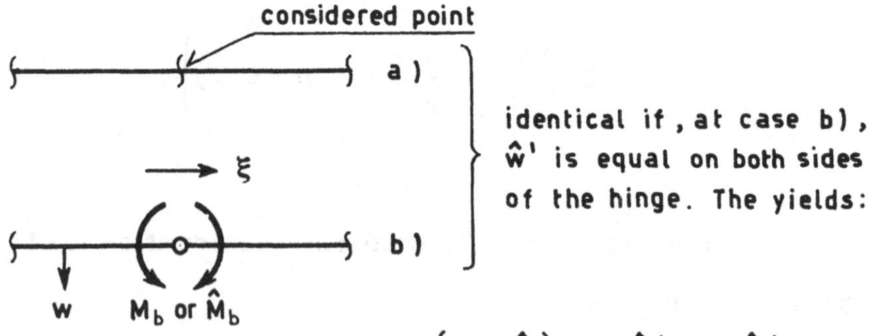

$$\Lambda_{ex}(M_b, \hat{w}) = M_b\hat{w}' - M_b\hat{w}' = 0\,.$$

Fig. 5. -6

2. The method of dummy displacements can be extend ed to more general cases. If, for example, we deal with a constant load per unit length and a constant EI, we know $w^{v} = 0$, or, if EI is a function $f(x)$ we can use w'' to be a linear function divided by $f(x)$.

6. CALCULUS OF VARIATIONS (cf. sect. 3.2 of the lecture).

6.1. A two-dimensional problem

Problem 6.1.-1 : Given a closed curve in space which can be projected on at least one plane in such a way that the projection line is a simply closed curve. Find out the surface to which this curve in space is the boundary and which has a minimal area. (cf. Fig. 6.1-1).

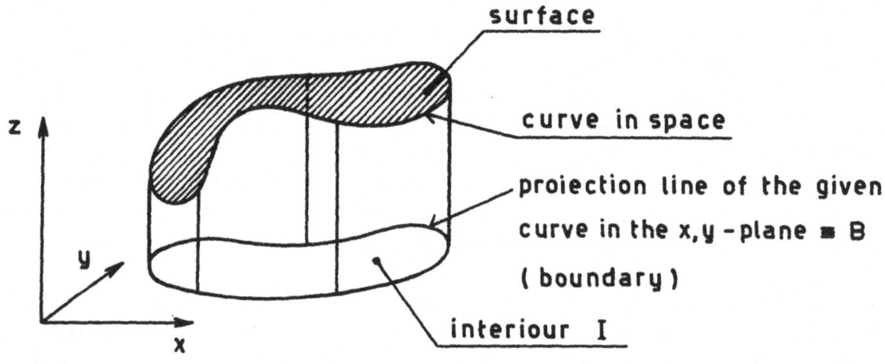

Fig. 6.1-1

First we have to express the surface and its area S in terms of two coordinates x, y as $z(x,y)$. The surface area will be denoted by S. The connection between dS and $dI \equiv$

$= dx\,dy$ can be seen from Fig. 6.1-2 :

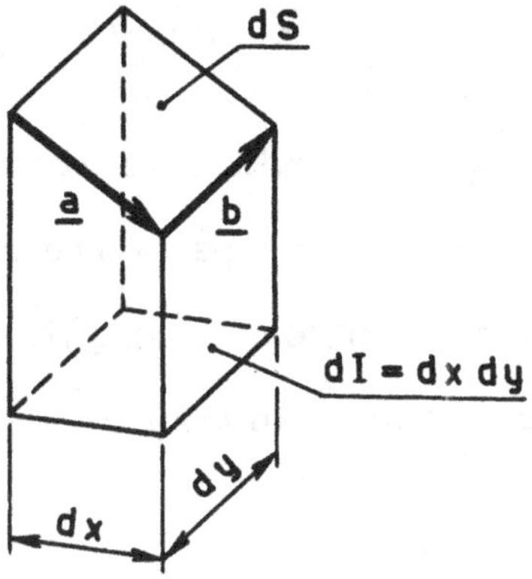

Fig. 6.1-2

$dS = |\underline{a} \times \underline{b}|$, where

$$\underline{a} = \begin{bmatrix} 1 \\ 0 \\ \frac{\partial z}{\partial x} \end{bmatrix} dx \quad ; \quad \underline{b} = \begin{bmatrix} 0 \\ 1 \\ \frac{\partial z}{\partial y} \end{bmatrix} dy$$

in an $\underline{e}_a$-system, where $a = x, y, z$.

So we get

$$dS = f\,dx\,dy \quad \text{with} \quad f = +\sqrt{1 + z_{,x}^2 + z_{,y}^2} \qquad \left(z_{,x} = \frac{\partial z}{\partial x}\ ;\ z_{,y} = \frac{\partial z}{\partial y} \right). \tag{1}$$

Therefore, our problem is stated

$$S = \iint^{I} f\,dx\,dy = \iint^{I} \sqrt{1 + z_{,x}^2 + z_{,y}^2}\,dx\,dy \Longrightarrow \min. \tag{2}$$

For the derivation of the equations let us assume that a more general problem is given :

$$S = \iint^{I} f\,(z_{,x}\,;\, z_{,y}\,;\, z\,;\, x\,;\, y)\; dx\, dy \quad \Longrightarrow \quad \min .$$

Furtheron, we will take into account that, possibly, parts of the boundary curve in space are not completely prescribed. We assume that $y(x)$ or $x(y)$ is given, but $z\,(x,y)$ is not given at parts of the boundary.

We introduce :

$$z = \tilde{z} + \varepsilon \zeta \,,$$

where $\tilde{z}$ is the desired solution, ζ is an arbitrary function and ε is a small parameter. Stationarity of the functional S in the neighbourhood of $\tilde{z}$ is necessary, thus we get

$$\frac{\partial S}{\partial \varepsilon} = 0$$

for any arbitrary function $\zeta(x,y)$. This function ζ has to be zero if $z\,(x,y)$ is prescribed (on the boundary). $\partial S / \partial \varepsilon = 0$ yields

$$\iint^{I} \left(\frac{\partial f}{\partial z_{,x}}\, \zeta_{,x} + \frac{\partial f}{\partial z_{,y}}\, \zeta_{,y} + \frac{\partial f}{\partial z}\, \zeta \right) dx\, dy = 0 \,,$$

or, using the GAUSS-GREEN-theorem :

$$\oint^{B} \left(\frac{\partial f}{\partial z_{,x}}\, \nu_x + \frac{\partial f}{\partial z_{,y}}\, \nu_y \right) \zeta\, ds + \int^{I} \left[\frac{\partial f}{\partial x} - \frac{\partial}{\partial x}\left(\frac{\partial f}{\partial z_{,x}} \right) - \frac{\partial}{\partial y}\left(\frac{\partial f}{\partial z_{,y}} \right) \right] \zeta\, dx dy,$$

where s is the arclength of the boundary B, and $\underline{\nu} = \nu_x \underline{e}_x + \nu_y \underline{e}_y$ is the unit vector normal to B and directed towards the outside of I.

First, we think of a LAGRANGE-variation. Then, we see from the fact that $\zeta(x,y)$ is arbitrary, that the integrand of $\int^{I} \ldots \zeta \, dx \, dy$ itself has to vanish :

$$\frac{\partial f}{\partial z} - \frac{\partial}{\partial x}\left(\frac{\partial f}{\partial z_{,x}}\right) - \frac{\partial}{\partial y}\left(\frac{\partial f}{\partial z_{,y}}\right) = 0 \ . \tag{3}$$

This has to be valid in the interior. On the boundary,

$$\frac{\partial f}{\partial z_{,x}} \nu_x + \frac{\partial f}{\partial z_{,y}} \nu_y = 0 \tag{4}$$

has to be given if $\zeta(x,y)$ is arbitrary, which means: on those parts of the boundary where z is not prescribed.

At this point we could have generalized the calculus to problems where the boundary itself is an arbitrary surface. But we will not do this here.

We specialize eqs. (3), (4) by using eq. (1) and get

$$\left(1 + z_{,y}^2\right) z_{,xx} + \left(1 + z_{,x}^2\right) z_{,yy} - 2 z_{,x} z_{,y} z_{,xy} = 0 \tag{5}$$

and

(6) $$z_{,x}\,\nu_x + z_{,y}\,\nu_y = 0 \qquad \text{(at a non prescribed boundary)}$$

Equation (5) can be interpreted as follows :

If the boundary curve in space is given, the problem is a physical problem and cannot depend on the coordinate system. Let us choose a system in which in the neighbourhood of a considered point, $z_{,x}$ and $z_{,y}$ vanish. This is always possible if the surface function is smooth enough. Then, we get at the considered point : $z_{,xx} + z_{,yy} = 0$. This means: the mean-curvature (not the GAUSS-curvature) of the surface which is normally called $\frac{1}{2}\,b_\alpha^\alpha$ in the theory of surfaces, has to vanish.

Equation (6) can be written as

$$\frac{\partial z}{\partial \xi} = 0 \,,$$

where ξ is the direction parallel to $\underline{\nu}$ in the x, y -plane. This means : S has a horizontal tangent in the direction of $\underline{\nu}$ at the boundary.

One example for minimal-area-surfaces are surfaces like the one printed in Fig. 6.1-3. (see next page).

If, in this example, $z\,(r, \vartheta)$ is prescribed at I and II, the described surface is the solution of the problem (1), even if z on III and IV is not prescribed. But if the curves III and IV are prescribed and z is arbitrary on I and II, the result will

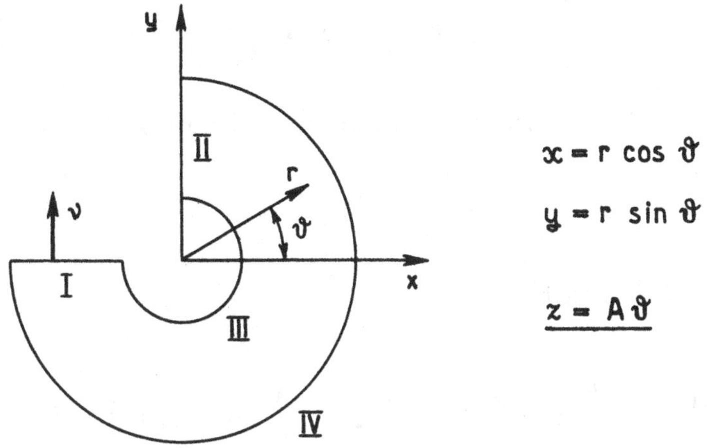

Fig. 6.1-3

not be $z = A\vartheta$, because this surface has not a horizontal tangent at I and II in the direction of $\underline{\nu}$.

6.2. A three-dimensional problem

Problem 6.2-1 : Assume that

$$U = \iiint\limits^{V} \frac{1}{2}\,\sigma_{ij}\,\varepsilon_{ij}\,dV \qquad (i,j \text{ here: cartesian system})$$

tends to become a minimum. Find out the EULER-LAGRANGE-equations for the case that U is expressed as functional of the displacements u_i (linear theory).

We start from the constitutive equations

$$\sigma_{ij} = 2\mu\,\varepsilon_{ij} + \lambda\,\delta_{ij}\,\varepsilon_{kk}$$

(μ, λ = LAME constants) and from the compatibility equation

$$\varepsilon_{ij} = \frac{1}{2}\,(u_{i,j} + u_{j,i}) .$$

This leads to

$$U = \iiint^{V} \frac{1}{2}\left[2\mu\,\varepsilon_{ij}\,\varepsilon_{ij} + \lambda\,\varepsilon_{ii}\,\varepsilon_{jj}\right] dV =$$

$$= \iiint \left[\frac{\mu}{4}\,(u_{i,j} + u_{j,i})(u_{i,j} + u_{j,i}) + \frac{\lambda}{2}\,u_{i,i}\,u_{j,j}\right] dV .$$

The problem is stated : $U \Longrightarrow \min$. We want to use only $U \Longrightarrow$ stationary or $\delta U = 0$ or $\dfrac{\partial U}{\partial \varepsilon} = 0$. We introduce

$$u_i = \tilde{u}_i + \varepsilon\bar{u}_i$$

($\tilde{u}_i$: solution, $\bar{u}_i$: arbitrary) and get

$$\iiint^{V} \left\{\frac{\mu}{2}\,(\tilde{u}_{i,j} + \tilde{u}_{j,i})(\bar{u}_{i,j} + \bar{u}_{j,i}) + \lambda\,\tilde{u}_{i,i}\,\bar{u}_{i,i}\right\} dV = 0 .$$

Partial integrations and the GAUSS-GREEN-theorem yield :

$$\oint^{B} \left\{ \frac{\mu}{2} (\tilde{u}_{i,j} + \tilde{u}_{j,i})(\bar{u}_i \nu_j + \bar{u}_j \nu_i) + \lambda \tilde{u}_{i,i} \bar{u}_j \nu_j \right\} dB -$$

$$- \iiint^{V} \left\{ \mu (\tilde{u}_{i,jj} + \tilde{u}_{j,ij}) + \lambda \tilde{u}_{j,ji} \right\} \bar{u}_i \, dI = 0 .$$

The boundary-integral is of no interest for us. The volume integral has to disappear for arbitrary functions $\bar{u}_i$. Thus, we can state

$$\iiint^{V} \left\{ \mu (\tilde{u}_{i,jj} + \tilde{u}_{j,ij}) + \lambda \tilde{u}_{j,ij} \right\} \bar{u}_i \, dI = 0$$

or

$$\underline{\underline{\mu \, u_{i,jj} + (\mu + \lambda) u_{j,ij} = 0 .}}$$

This very well-known result is normally derived in the following manner :

$$\sigma_{ij} = \mu (u_{i,j} + u_{j,i}) + \lambda \delta_{ij} u_{k,k}$$

$\sigma_{ij,i} = 0$ then yields

$$\mu (u_{i,ji} + u_{j,ii}) + \lambda \delta_{ij} u_{k,ki} =$$

$$= \mu (\mu_{j,ii} + \mu_{i,ji}) + \lambda u_{i,ij} =$$

$$= \mu \, u_{j,ii} + (\mu + \lambda) u_{i,ji} = 0 .$$

7. LAGRANGE EQUATIONS OF SECOND KIND (cf. sect. 3. 4. 2 of the lecture).

The LAGRANGE-equations of the 2nd kind are very well known. Therefore, only a special question will be illustrated by an example.

Problem 7. -1 : Find out the equations of motion for the problem sketched in Fig. 7. -1 by use of the coordinates

a) $x, \xi, \eta,$ and

b) x, ψ.

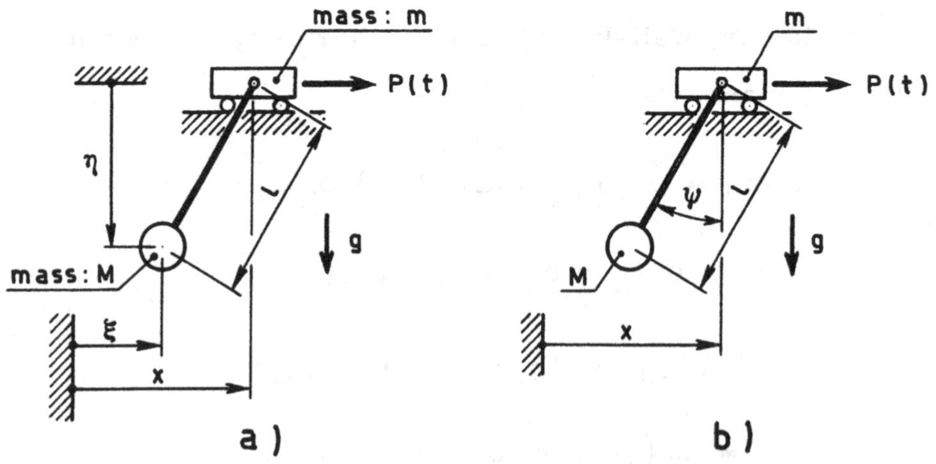

Fig. 7. -1

a) We use three coordinates despite the fact that we have only two degrees of freedom. Thus, we have to deal with one constraint :

$$g = (\xi - x)^2 + \eta^2 - \ell^2 = 0 . \tag{1}$$

The kinetic energy is given by

$$T = \frac{1}{2} m\dot{x}^2 + \frac{1}{2} M\dot{\xi}^2 + \frac{1}{2} M\dot{\eta}^2 ,$$

whereas the potential energy is

$$U = - Mg\eta \quad (\eta = 0 \; : \; U = 0).$$

The modified kinetic potential L^P is then

$$L^P = \frac{1}{2} m\dot{x}^2 + \frac{1}{2} M\dot{\xi}^2 + \frac{1}{2} M\dot{\eta}^2 + Mg\eta + \lambda\left[(\xi - x)^2 + \eta^2 - \ell^2\right] .$$

The right sides of the equations of motion are calculated by means of the virtual work theorem :

$$\delta' W^r = \sum_j Q^r_j \, \delta' x_j \Longrightarrow P\,\delta' x = Q^r_x \, \delta' x + Q^r_\xi \, \delta'\xi + Q^r_\eta \delta'\eta .$$

This can be interpreted as :

case α) $\qquad Q^r_x = P , \quad Q^r_\xi = Q^r_\eta = 0.$

But we can also use

$$\delta' x = \frac{\eta}{\xi - x}\,\delta'\eta + \delta'\xi$$

or

$$P\,\delta' x = P\,\frac{\eta}{\xi - x}\,\delta'\eta + P\,\delta'\xi$$

Hence,

case β) $\quad Q^r_x = 0\,, \quad Q^r_\xi = P\,, \quad Q^r_\eta = \dfrac{\eta}{\xi - x}\,P.$

The equation

$$\frac{d}{dt}\left(\frac{\partial L^P}{\partial \dot q_j}\right) - \frac{\partial L^P}{\partial q_j} = Q^r_j$$

leads to

			case α)		case β)
(2)	$m\ddot x - 2\lambda(x - \xi)$	$=$	P	$=$	$0\,,$
(3)	$M\ddot\xi - 2\lambda(\xi - x)$	$=$	0	$=$	$P\,,$
(4)	$M\ddot\eta - Mg - 2\lambda\eta$	$=$	0	$=$	$\dfrac{\eta}{\xi - x}\,P\,.$

Equation (2) allows us to calculate 2λ as

case α) $2\lambda = \frac{1}{x-\xi}(m\ddot{x} - P)$; case β) $2\lambda = \frac{m\ddot{x}}{x-\xi}$.

Equations (3), (4) then read

case α) case β)

$$M\ddot{\xi} + m\ddot{x} - P = 0 \quad \text{or} \quad M\ddot{\xi} + m\ddot{x} = P \tag{5}$$

$$M\ddot{\eta} - Mg + \frac{\eta}{\xi - x}(m\ddot{x} - P) = 0 \quad \text{or} \quad M\ddot{\eta} - Mg + \frac{\eta}{\xi - x} m\ddot{x} = \frac{\eta}{\xi - x} P . \tag{6}$$

We see : If we use LAGRANGE-equations of the 2nd kind at problems with constraints, the Q_j^r are not uniquely determinate. But the equations of motion derived by this meth od are identical in every case. The value of λ, only, depends on Q_j^r.

b) For comparison, we see here the usual solution of this problem :

$$T = \frac{1}{2}(m + M)\dot{x}^2 - M\dot{x}\dot{\psi}\ell\cos\psi + \frac{1}{2}M\ell^2\dot{\psi}^2 ,$$

$$U = -Mg\ell\cos\psi ,$$

$$L = T - U ,$$

$$\delta W^r = P\,\delta' x = Q^r_x\,\delta' x + Q^r_\psi\,\delta'\psi\ ,$$

hence

$$Q^r_x = P,\ Q^r_\psi = 0.$$

$$\frac{d}{dt}\left(\frac{\partial L}{\partial \dot{q}_j}\right) - \frac{\partial L}{\partial q_j} = Q^r_j \quad \text{now yields}$$

(7) $$(m + M)\ddot{x} - M\ell\ddot{\psi}\cos\psi + M\ell\dot{\psi}^2\sin\psi = P,$$

$$- M\ddot{x}\ell\cos\psi + M\cancel{\dot{x}\ell\dot{\psi}}\sin\psi + M\ell^2\ddot{\psi} -$$

(8) $$- M\cancel{\dot{x}\dot{\psi}\ell}\sin\psi + Mg\ell\sin\psi = 0\ .$$

To compare this result with eqs. (5) and (6) we introduce

$$\xi = x - \ell\sin\psi\ ;\quad \dot{\xi} = \dot{x} - \ell\dot{\psi}\cos\psi\ ;\quad \ddot{\xi} = \ddot{x} - \ell\ddot{\psi}\cos\psi + \ell\dot{\psi}^2\sin\psi,$$

$$\eta = \ell\cos\psi\ ,\quad \dot{\eta} = -\ell\dot{\psi}\sin\psi\ ;\quad \ddot{\eta} = -\ell\ddot{\psi}\sin\psi - \ell\dot{\psi}^2\cos\psi.$$

Equation (7) reads

$$m\ddot{x} + M(\ddot{x} - \ell\ddot{\psi}\cos\psi + \ell\dot{\psi}^2\sin\psi) = P \quad \text{or}$$

$$m\ddot{x} + M\ddot{\xi} = P \,. \tag{9}$$

This is identical with eq. (5).

To compare eqs. (6) and (8), we start from eq. (6) :

$$-M(\ell\ddot{\psi}\sin\psi + \ell\dot{\psi}^2\cos\psi) - Mg - \frac{\ell\cos\psi}{\ell\sin\psi}(m\ddot{x} - P) = 0$$

or using eq. (9) :

$$-M\ell\ddot{\psi}\sin\psi - M\ell\dot{\psi}^2\cos\psi - Mg + \operatorname{cotan}\psi\, M\ddot{\xi} = 0$$

or multiplying by $\ell\sin\psi$ and inserting $\ddot{\xi}$:

$$-M\ell^2\ddot{\psi}\sin^2\psi - M\ell^2\cancel{\dot{\psi}^2\sin\psi}\cos\psi - Mg\ell\sin\psi +$$

$$+ M\ell\cos\psi\,(\ddot{x} - \ell\ddot{\psi}\cos\psi + \cancel{\ell\dot{\psi}^2\sin\psi}) = 0\,,$$

$$-M\ell^2\ddot{\psi} - Mg\ell\sin\psi + M\ell\ddot{x}\cos\psi = 0\,.$$

This equation is identical with eq. (8).

8. RAYLEIGH RITZ METHOD APPLIED TO EIGENFREQUENCIES OF ONE DIMENSIONAL CONTINUOUS SYSTEM (cf. sect. 3. 4. 3 of the lecture).

In this chapter, two problems have to explain the method and to show the convergency of the solution if at least two free parameters are used.

Problem 8. -1 : Find out approximations of the first eigenfrequencies belonging to symmetric eigenmodes for the simply supported straight beam of Fig. 8. -1, by applying the RAYLEIGH-RITZ-method.

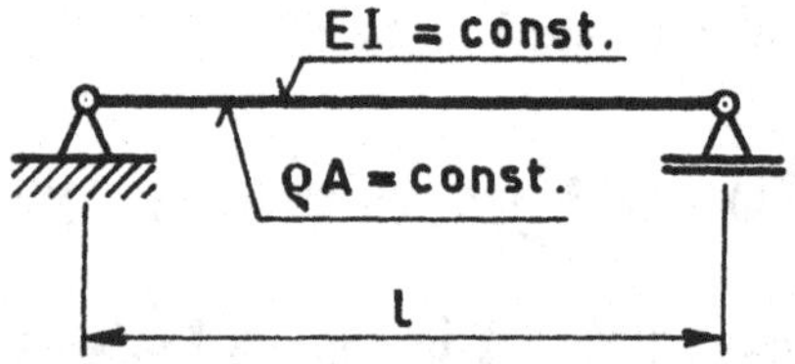

Fig. 8. -1

The exact solution for symmetric eigenmodes is

$$\omega = b\pi^2\sqrt{\frac{EI}{\rho A l^4}} \qquad \text{with} \qquad b = 1^2, 3^2, 5^2, \ldots$$

For the approximation of the symmetric eigenmodes we can use the following series of modes (cf. Fig. 8. -2) :

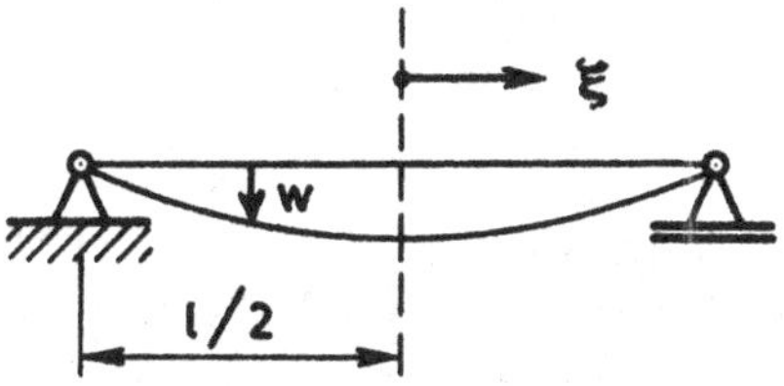

Fig. 8.-2

$$w_1 = 1 - \left(\frac{2\xi}{\ell}\right)^2 \quad ; \qquad w_1'' = -2 \cdot 1 \cdot \frac{4}{\ell^2}$$

$$w_2 = 1 - \left(\frac{2\xi}{\ell}\right)^4 \quad ; \qquad w_2'' = -4 \cdot 3 \cdot \frac{4}{\ell^2}\left(\frac{2\xi}{\ell}\right)^2$$

$$w_n = 1 - \left(\frac{2\xi}{\ell}\right)^n \quad ; \qquad w'' = -n\,(n-1)\,\frac{4}{\ell^2}\left(\frac{2\xi}{\ell}\right)^{2n-2}$$

The kinetic energy is

$$T(\dot{a}) = \frac{1}{2}\int \rho A \dot{w}^2 \, d\xi + \ldots .$$

The dots (. . .) denote that additional terms (point-masses etc.) may appear. In the RAYLEIGH-RITZ-method we use

$$T(a) = \frac{1}{2}\int \rho F \, w^2 d\xi + \ldots$$

and

$$U(a) = \frac{1}{2}\int E\,I\,(w'')^2 d\xi + \ldots$$

We know that the smallest eigenvalue ω_1 satisfies the inequality

$$\omega_1^2 = \min\left(\frac{U(a)}{T(a)}\right) \leq \min\left(\frac{U^*(a^*)}{T^*(a^*)}\right),$$

where $U^*(a^*)$ and $T^*(a^*)$ are $U(a)$ and $T(a)$ in restricted spaces $a = a^*$. We introduce

$$w^* = \sum_i \alpha_i w_i .$$

So we get

$$T^*(a^*) = \frac{1}{2}\int \varrho A w^{*2} d\xi + \ldots =$$

$$= \frac{1}{2}\int \varrho A \sum_{i,j} w_i^* w_j^* \alpha_i \alpha_j d\xi + \ldots =$$

$$= \sum_{i,j} \alpha_i \alpha_j \left(\frac{1}{2}\int \varrho A w_i^* w_j^* d\xi + \ldots\right) \overset{!}{=} \sum_{i,j} \alpha_i \alpha_j T_{ij} .$$

Correspondingly, we get

$$U^*(a^*) = \sum_{i,j} \alpha_i \alpha_j U_{ij} , \quad \text{where}$$

$$T_{ij} = \frac{1}{2} \quad \varrho A w_i^* w_j^* d\xi ,$$

$$U_{ij} = \frac{1}{2} \quad E I w_i^{*''} w_j^{*''} d\xi .$$

This leads to

$$\omega_1^2 \le \min \left(\frac{\sum\limits_{i,j} \alpha_i U_{ij} \alpha_j}{\sum\limits_{i,j} \alpha_i T_{ij} \alpha_j} \right) .$$

Comparing this with sect. 1. 3, we recognize that this problem belongs to

$$\omega_1^2 \le \omega^2 \quad \text{with} \quad \sum_j \left(\omega^2 T_{ij} - U_{ij}\right) \alpha_j = 0$$

which yields

$$\underline{\underline{\det \left(\omega^2 T_{ij} - U_{ij}\right) = 0}} \; .$$

In our example the values of T_{ij} and U_{ij} are :

$$T_{11} = \frac{4}{15} \rho A \ell \qquad U_{11} = 32 \; EI/\ell^3 ,$$

$$T_{12} = \frac{32}{105} \rho A \ell \qquad U_{12} = 64 \; EI/\ell^3 ,$$

$$T_{22} = \frac{16}{45} \rho A \ell \qquad U_{22} = \frac{1152}{5} EI/\ell^3 .$$

<u>1. approximation</u>

To compute a first approximation of ω_1^2 we use only w_1: $\det\left(\omega^2 T_{11} - U_{11}\right) = 0 = \omega^2 T_{11} - U_{11}$ leads to

$$\omega_1^2 \le \frac{U_{11}}{T_{11}} = 120 \frac{EI}{\rho A \ell^4} \quad \text{or} \quad \omega_1 \le 10.96 \sqrt{\frac{EI}{\rho A \ell^4}} ,$$

whereas the correct result is $\omega_1 = 9.87 \sqrt{\dfrac{EI}{\rho A \ell^4}}$.

2. approximation

We use w_1 and w_2 . Then we get with $\omega = \bar{\omega}\sqrt{\frac{EI}{\rho A l^4}}$

$$\left| \bar{\omega}^2 \begin{bmatrix} \frac{4}{15} & \frac{35}{105} \\ \frac{32}{105} & \frac{16}{45} \end{bmatrix} - \begin{bmatrix} 32 & 64 \\ 64 & \frac{1152}{5} \end{bmatrix} \right| = 0$$

or

$$\bar{\omega} = \begin{cases} 9.87 \\ 131.7 \end{cases} .$$

We will not derive here that the method yields as its second result an upper bound for the second eigenfrequency if the first eigenvalue is approximated very well. In our example we see : The first eigenvalue ($\bar{\omega} = 9.87$) is approximated very exactly. Therefore, $\bar{\omega} = 131.7$ should be an upper bound to $\bar{\omega}_2$. Indeed, it is a (bad) upper bound to the true value which is 88.8 .

Problem 8.- 2 : Find out approximate solutions for the first two eigenfrequencies of the frame sketched in Fig. 8.- 3. Apply a linear theory and neglect rotational terms. Deal with pure bending (no shear deformations etc.).

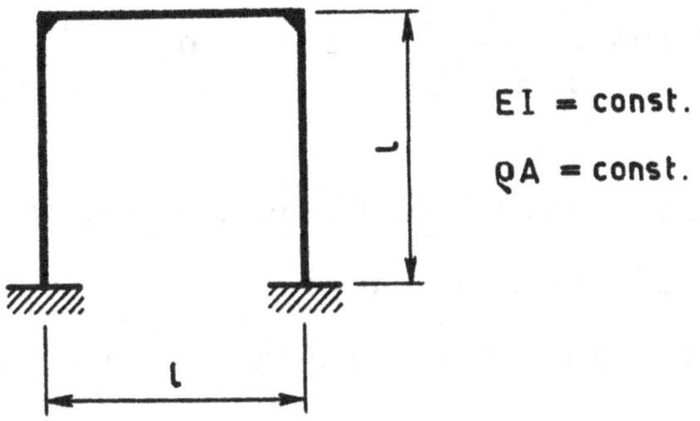

Fig. 8. -3

The restriction to a linear theory means that terms induced by tension-forces have to be neglected. They would yield quadratic terms (cf. Fig. 8. -4).

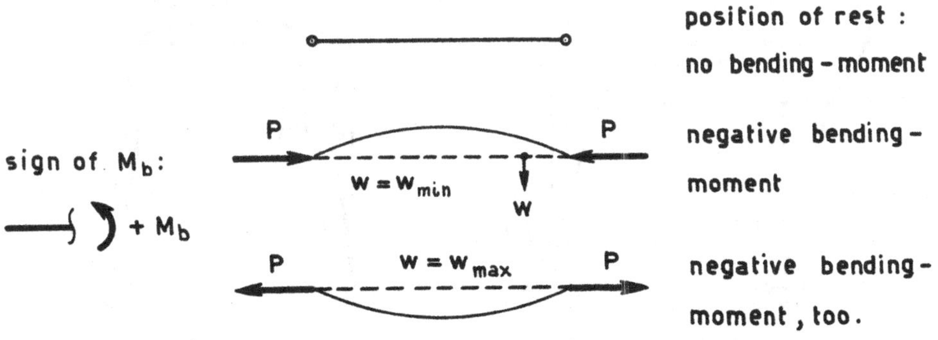

Fig. 8. -4

The first two eigenfrequencies of the frame are correctly :

$$\omega_1 = 3.204\sqrt{\frac{EI}{\rho A \ell^4}} \quad \text{and} \quad \omega_2 = 12.7\sqrt{\frac{EI}{\rho A \ell^4}} \; .$$

The computation of these values is a problem for itself but it is of no interest here.

We can imagine that there must exist symmetric and antisymmetric eigenmodes like those ones sketched in Fig. 8.-5. Therefore, we can apply the approximation method to calculate both frequencies independently in subspaces of anti-symmetric and symmetric eigenmodes. Then, the upper bound theorem remains valid also for the second eigenfrequency. ω_1^2 and ω_2^2 belong to real minima of the RAYLEIGH-quotient.

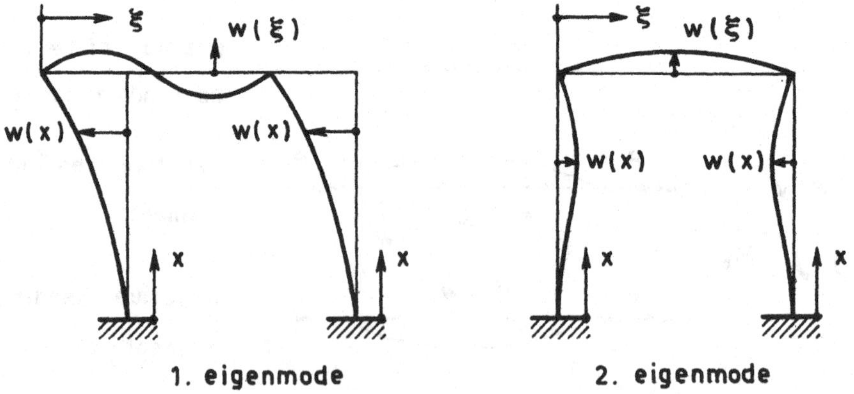

Fig. 8.-5

We start with the second eigenmode because it is simpler for the calculation.

First approximation for the second eigenmode

The coordinates used at both eigenmodes are printed in Fig. 8.-5.

$w(x)$ can be approximated by

$$w(x) = a\left(x^3 - \ell x^2\right) \qquad \text{(third order)}$$

whereas $w(\xi)$ is given by

$$w(\xi) = a\ell\ \xi(\ell - \xi) \qquad \text{(second order, symmetric to } \xi = \tfrac{\ell}{2}\text{)}$$

The geometrical boundary conditions

$$w(x=0) = w(x=\ell) = w'(x=0) =$$

$$= w(\xi=0) = w(\xi=\ell) = 0$$

and

$$w'(x=\ell) = w'(\xi=0)$$

are satisfied as we can check by inserting $w(\xi)$ and $w(x)$.

$$\omega_2^2 \le \frac{U\left(a^{\text{symmetric}}\right)}{T\left(a^{\text{symmetric}}\right)}$$

leads to

$$\omega_2^2 \le \frac{EI}{\rho A} \cdot \frac{2\int_0^\ell \left[w''(x)\right]^2 dx + \int_0^\ell \left[w''(\xi)\right]^2 d\xi}{2\int_0^\ell \left[w(x)\right]^2 dx + \int_0^\ell \left[w(\xi)\right]^2 d\xi} .$$

With $w(x)$ and $w(\xi)$ as they are estimated and with $w''(x) =$

$$w''(x) = a\,(6x - 2\ell)\,, \quad w''(\xi) = -2a\ell$$

we get

$$\omega_2^2 \leq 229.1\,\frac{EI}{\rho A\ell^4} \qquad \text{or} \qquad \omega_2 \leq 15.1\sqrt{\frac{EI}{\rho A\ell^4}}$$

so that the error is 18.9 % .

Second approximation for the second eigenmode

A mistake can be expected especially from $w(\xi)$ which was approximated by a second-order polynomial. Hence, we use now

$$w(x) = a\,(x^3 - \ell x^2) \quad \text{and}$$

$$w(\xi) = a\left[\ell\xi(\ell - \xi) + \alpha\,(\ell^3\xi^2 - 2\ell\xi^3 + \xi^4)\right],$$

where the additional term $\ell^2\xi^2 - 2\ell\xi^3 + \xi^4$ has the form, printed in Fig. 8.-6 and does not enter the boundary condition.

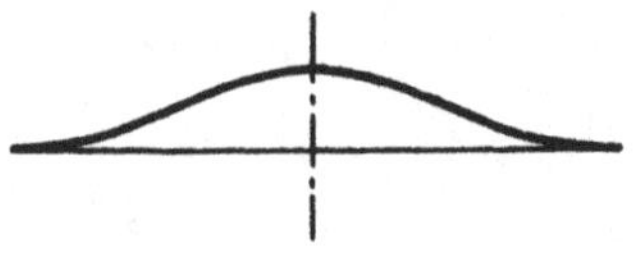

Fig. 8.-6

We calculate a new result from eq. (1) :

$$\omega_2^2 \leq 504 \frac{EI}{\varrho A \ell^4} g(\alpha\ell) \quad \text{where} \quad g(\alpha\ell) = \frac{15 + (\alpha\ell)^2}{33 + 9(\alpha\ell) + (\alpha\ell)^2} .$$

Optimization (minimization) of $g(\alpha\ell)$ yields $\alpha\ell = \sqrt{19} - 2$ and

$$\omega_2^2 \leq 173 \frac{EI}{\varrho A \ell^4} \quad \text{or} \quad \omega_2 \leq 13.15 \sqrt{\frac{EI}{\varrho A \ell^4}} ,$$

so that the error is reduced to 3.5 % .

First approximation for the first eigenmode

Looking at the left side of Fig 8.-5, we see that the simplest way to describe an antisymmetric eigenmode is

$$w(x) = a x^2 \; ; \qquad w(\xi) = 2 a \ell \left(\xi - 3 \frac{\xi^2}{\ell} + 2 \frac{\xi^3}{\ell^2} \right),$$

where the following boundary conditions are satisfied :

$$w(x=0) = w'(x=0) = w(\xi=0) = w\left(\xi = \frac{\ell}{2}\right) = w(\xi=\ell) = 0,$$

$$w'(\xi=0) = w'(x=\ell) = w'(\xi=\ell) \text{ and :}$$

antisymmetry of $w(\xi)$ with respect to $\xi = \frac{\ell}{2}$.

Instead of formula (1) we get now

$$\omega_1^2 \leq \frac{EI}{\varrho A} \quad \frac{2\int_0^{\ell}[w''(x)]^2 dx + \int_0^{\ell}[w''(\xi)]^2 d\xi}{2\int_0^{\ell}[w(x)]^2 dx + \int_0^{\ell}[w(\xi)]^2 d\xi + \ell[w(x=\ell)]^2} \tag{2}$$

The additional term $\varrho A \ell [w(x=\ell)]^2$ in the denominator takes care of the fact that the horizontal beam vibrates in the horizontal direction like a point-mass.

The first approximation of the first eigenmode yields by means of eq. (2) :

$$\omega_1^2 \leq 39.4 \frac{EI}{\varrho A \ell^4} \quad \text{or} \quad \omega_1 \leq 6.28 \sqrt{\frac{EI}{\varrho A \ell^4}} ,$$

so that the error is 96 % of the true value. This result lies not even in the neighbourhood of the correct solution.

<u>Second approximation for the first eigenmode</u>

We can assume that the main reason for the bad result of the first approximation is that we used a polynomial of second order in the vertical parts of the frame. Therefore, we start from

$$w(x) = a\left[(1+\alpha)\ell x^2 - a x^3\right] ;$$

$$w(\xi) = (2-\alpha)a\left(\ell^2\xi - 3\ell\xi^2 + 2\xi^3\right)$$

and get with an optimal value of α of $\alpha_{opt} = 1.45$:

$$\omega_1 = 3.215 \sqrt{\frac{EI}{\rho A l^4}}$$

which represents an error of only 0.3 %

Discussion of the results.

At the true solution a further (non-geometrical) transition condition has to be satisfied at $x = l$, $\xi = 0$: The bending-moment which is represented by w'' because of $EI =$ =const., must be the same on both sides of this point. Thus, we would have to satisfy

$$w''(x = l) \overset{!}{=} w''(\xi = 0). \tag{3}$$

This relation is not even approximately satisfied in our first approximations. Looking at Fig 8.-5, we immediately see that in both cases the curvature w'' jumps from positive to negative values at this point in the first approximations. Possibly, the optimization process of the second approximations leads to a smaller error of eq. (3). With $\alpha = \alpha_{opt}$ we get :

2. eigenmode : 1st approximation $4al \longleftrightarrow -2al$ (bad)
   2nd approximation $4al \longleftrightarrow +2{,}8al$ (better)

1. eigenmode : 1st approximation : $2a \longleftrightarrow -12a$ (bad)

: 2nd approximation :

$$-3{,}8\ a\ell \longleftrightarrow -3{,}3\ a\ell \qquad \text{(nearly correct)}$$

Let us discuss a further question :

By what reason is the second approximation of the first eigenmode so good?

A very important part of the kinetic energy of the beam belongs to the horizontal vibration of the horizontal beam. Therefore, the corresponding inertia force is very large. Hence, we have relatively high values of $|w''(x)|$ in the vertical beams (high bending moment) but there is no reason for large deflections in the horizontal beam. This is not compatible with our first assumption : $w(x) = a x^2$. This is the reason for the insufficient first approximation and for the extremely good result at the second approximation : M_b is nearly linear (w is cubic, as it is estimated) in the vertical beam, whereas the bending-moment in the horizontal beam is less important. By this reason an approximation in which the horizontal beam is taken to be rigid $\left(w(\xi) \equiv 0\right)$ leads to a better result than our "first approximation" :

$$\omega_1^2 \leq 13.8\ \frac{EI}{\rho A \ell^4} \quad \text{(true: } 10.34\ /\!/\ \text{1st approximation: } 39.4).$$

Even if we drop the mass in the vertical beams and take $w(\xi) \equiv 0 \quad \left(w'(x=\ell) = 0\right)$ the eigenfrequency of the resulting system of one degree of freedom (Fig. 8. -7) is smaller than the first approximation :

$$\omega_1^2 \leq 24 \frac{EI}{\varrho A \ell^4}$$

It is to be remarked that dropping of masses decreases the denominator and increases the upper bound. So it is allowed.

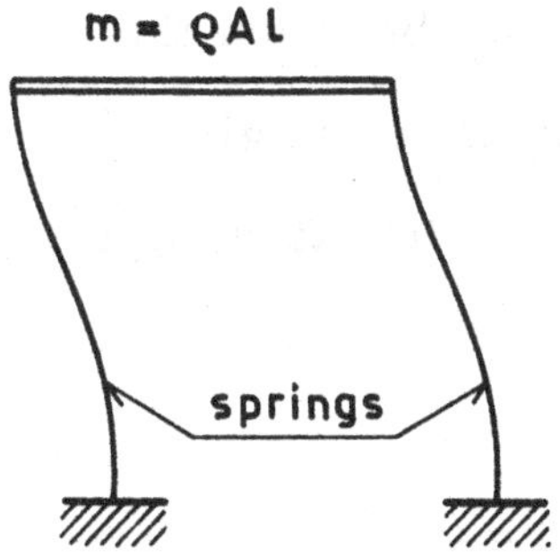

Fig. 8. -7

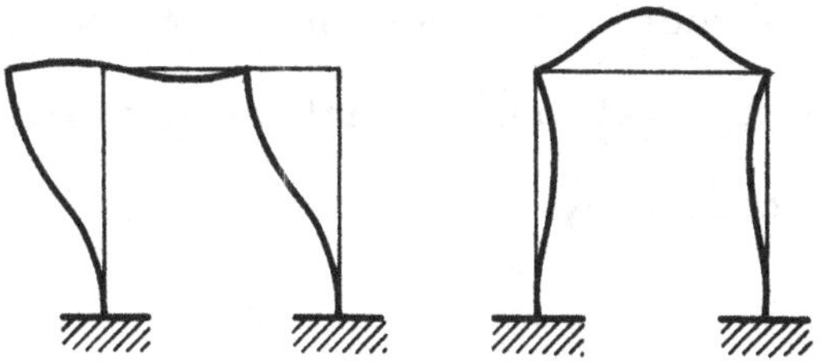

Fig 8. -8

As a final remark the correct modes are sketched in Fig. 8. -8. They are quite different from the modes of Fig. 8. -5.

9. HARMONIC BALANCE (cf. sect. 3. 4. 3. of the lecture).

In this chapter, the method of harmonic balance will be applied to two relatively simple problems of nonlinear vibrations.

Problem 9. -1 : Find out approximate relations between the amplitude of the response A, p, ω, and the phase angle φ for the stationary solution of a problem with nonlinear damping described by

$$\mathcal{L} \equiv m\ddot{q} + a\dot{q}^3 + cq - p\cos\omega t = 0 .$$

$\mathcal{L}$ is treated to be the EULER-LAGRANGE-equation of a HAMILTON's problem :

$$\int_{t_1}^{t_2} \mathcal{L}\eta \, dt - \Big[\eta\, f(\mathcal{L})\Big]_{t_1}^{t_2} = 0 .$$

The period T of the response is prescribed by the external force $p\cos\omega t$ as to be $T = 2\pi/\omega$. We can assume that η is periodic with the period $2\pi/\omega$. Then, $f(\mathcal{L})$ has the period $T = 2\pi/\omega$, too. Thus, we get :

$$\Big[\eta\, f(\mathcal{L})\Big]_{t_1}^{t_1+T} = \big(\eta\, f(\mathcal{L})\big)\Big|_{t_1+T} - \big(\eta\, f(\mathcal{L})\big)\Big|_{t_1} = 0 .$$

This yields

$$\int_{t_1}^{t_1+T} \mathcal{L}\,\eta\,dt = 0.$$

We search for an approximate solution $q(t)$ to the original problem in the subspace of

$$q(t) = A\cos(\omega t + \varphi).$$

A and φ are variable, ω is a given value. Hence, $\delta q(t)$ can be completely described by

$$\delta' q(t) = \delta' A \cos(\omega t + \varphi) - A\,\delta'\varphi \sin(\omega t + \varphi),$$

where $\delta' A$ and $\delta'\varphi$ are arbitrary values. So η can be $\cos(\omega t + \varphi)$ or $\sin(\omega t + \varphi)$ only, the constant coefficients do not matter.

For simplicity, we introduce

$$\omega t + \varphi = \lambda \quad \text{or} \quad \omega t = \lambda - \varphi .$$

Then,

$$\int_{t_1}^{t_1+T} \mathcal{L}\,\eta\,dt = 0$$

leads to

$$\int_0^{2\pi} \mathcal{L}(\lambda)\,\eta(\lambda)\,d\lambda = 0.$$

The function $\mathcal{L}(\lambda)$ is, in our problem,

$$\mathcal{L}(\lambda) = -m\omega^2 A\cos\lambda -$$

$$-a\omega^3 A^3\sin^3\lambda + cA\cos\lambda - p\cos(\lambda-\varphi),$$

whereas $\eta(\lambda)$ is equal to $\sin\lambda$ or $\cos\lambda$.

Using

$$\int_0^{2\pi}\sin^n\lambda\cos\lambda\,d\lambda = \int_0^{2\pi}\cos^n\lambda\sin\lambda\,d\lambda = 0,$$

$$\int_0^{2\pi}\sin^2\lambda\,d\lambda = \int_0^{2\pi}\cos^2\lambda\,d\lambda = \pi,$$

$$\int_0^{2\pi}\sin^4\lambda\,d\lambda = \int_0^{2\pi}\cos^4\lambda\,d\lambda = \frac{3}{4}\pi,$$

$$\cos(\lambda-\varphi) = \cos\lambda\cos\varphi + \sin\lambda\sin\varphi,$$

we get :

<u>if $\eta = \sin\lambda$</u> : $\quad 0 - a\omega^3 A^3\frac{3}{4}\pi + 0 - p\sin\varphi\,\pi = 0,$

<u>if $\eta = \cos\lambda$</u> : $\quad -m\omega^2 A\pi - 0 + cA\pi - p\cos\varphi\,\pi = 0$

or

$$p \sin\varphi = -\frac{3}{4} a A^3 \omega^3 \; ; \quad p \cos\varphi = A(c - m\omega^2).$$

Now it is easy to express p and φ as functions of A and ω :

$$p = mA \overset{+}{\sqrt{\left(\frac{c}{m} - \omega^2\right)^2 + \frac{9}{16}\left(\frac{a}{m}\right)^2 A^4 \omega^6}} \, ,$$

$$\operatorname{sign}(\sin\varphi) = -\operatorname{sign}(a) \, ,$$

$$\cos\varphi = \left(\frac{c}{m} - \omega^2\right) \cdot \left[\left(\frac{c}{m} - \omega^2\right)^2 + \frac{9}{16}\left(\frac{a}{m}\right)^2 A^4 \omega^6\right]^{-\frac{1}{2}}.$$

The nonlinearity of the damping appears in the terms $\frac{9}{16}\left(\frac{a}{m}\right)^2 A^4 \omega^6$. But it is quite funny that there is a solution if $a < 0$. For such a negative damping the stationary solution is unstable Nevertheless, it is approximated by the method of harmonic balance.

<u>Problem 9. -2</u> : Find out an approximate solution of the type

$$q(t) = q_0 + A \cos(\omega t + \varphi)$$

for a problem which can be described by

$$m\ddot{q} + cq(1 + \alpha q) = p \cos \omega t \, .$$

In the case of problem 9. -2 $\mathcal{L}$ is equal to

$$\mathcal{L} = m\ddot{q} + cq(1 + \alpha q) - p\cos\omega t\,.$$

The variation of q in the subspace of $q = q_0 + A\cos(\omega t + \varphi)$ is to be expressed by

$$\delta' q = \delta' q_0 + \delta' A\cos(\omega t + \varphi) - A\delta'\varphi\sin(\omega t + \varphi);$$

$\delta' q_0$, $\delta' A$ and $\delta'\varphi$ are arbitrary, therefore, η can be :

$$\eta^1 = 1\,,\quad \eta^2 = \cos(\omega t + \varphi)\,,\quad \eta^3 = \sin(\omega t + \varphi);$$

We introduce (like at problem 9.-1) $\lambda = \omega t + \varphi$.

Thus, we get

$$\mathcal{L} = -mA\omega^2\cos\lambda + c(q_0 + A\cos\lambda)(1 + \alpha q_0 + \alpha A\cos\lambda) - p\cos(\lambda - \varphi)$$

or

$$\mathcal{L} = cq_0(1 + \alpha q_0) + \left[-mA\omega^2 + cA(1 + 2\alpha q_0) - p\cos\varphi\right]\cos\lambda + \\ + c\alpha A^2\cos^2\lambda - p\sin\varphi\sin\lambda$$

and

$$\eta^1 = 1\,,\quad \eta^2 = \cos\lambda\,,\quad \eta^3 = \sin\lambda\,.$$

Now, $\int_0^{2\pi} \mathcal{L}\eta\, d\eta = 0$ yields :

(1) $\eta = \eta^1$: $\quad 2\pi c q_0(1 + \alpha q_0) + \pi c\alpha A^2 = 0\,,$

(2) $\eta = \eta^2$: $\quad \pi\left[-mA\omega^2 + cA(1 + 2\alpha q_0) - p\cos\varphi = 0\,,\right.$

(3) $\eta = \eta^3$: $\quad -\pi p\sin\varphi = 0\,.$

We try to express q_0, φ, and p as functions of A and ω. First, we derive simply from eq. (3)

$$\sin \varphi = 0$$

($p = 0$ would not have any sense). We define: $\varphi = 0$, so we exclude $\varphi = \pi$ etc. This has some influence on the sign of p etc. From $\varphi = 0$ we get $\cos \varphi = 1$ and with eq. (2) :

$$p = A\left[c\,(1 + 2\alpha q_0) - m\omega^2\right].$$

Equation (1) gives us q_0 as a function of A and ω :

$$q_0 = -\frac{1}{2\alpha}\left[1 - \sqrt{1 - 2\,(\alpha A)^2}\right]. \tag{4}$$

This is a first result. For small values of α we can write :

$$q_0 \approx -\frac{1}{2\alpha}\left[1 - (1 - \alpha^2 A^2)\right] = -\frac{\alpha^2 A^2}{2\alpha} = -\frac{\alpha A^2}{2}. \tag{5}$$

Inserting q_0 given by eq. (4) into the equation for p, we get

$$p = A\left[c\sqrt{1 - 2\,(\alpha A)^2} - m\omega^2\right]. \tag{6}$$

We see that for $\alpha \to 0$ (linear case) we get the well-known results

$$\varphi = 0 , \quad q_0 = 0 , \quad p = A(c - m\omega^2).$$

The most interesting effect is that we get a term q_0. At problem 9.-1 it would have been zero. This effect is typical for nonlinear problems. Let us interprete it physically by means of Fig. 9.-1.

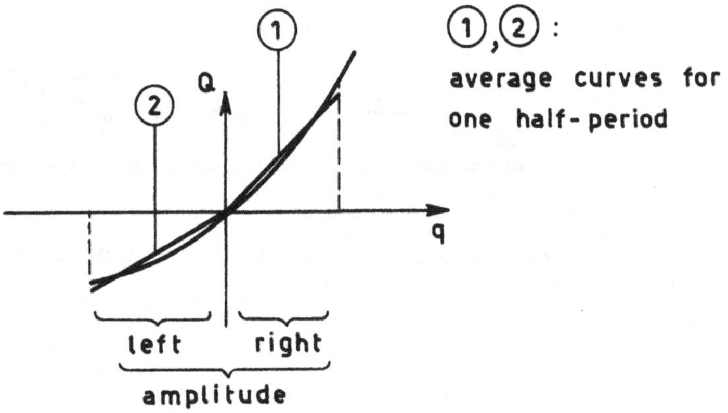

Fig. 9.-1

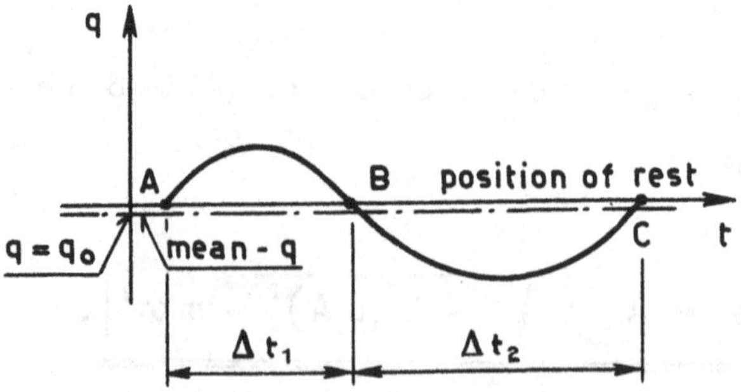

Fig. 9.-2

For the interpretation we think of $p = 0$. Then, the only acting force is given by $Q(q) = cq(1 + \alpha q)$. This is $(\alpha > 0)$ presented at the diagram (Fig. 9.-1). The force can - for one half-period - be approximated by any average straight line passing the point $q = Q = 0$, for $q > 0$ by curve ①, for $q < 0$ by curve ②. These curves can be taken as spring-force curves. We see that the spring stiffness for $q > 0$ is higher than for $q < 0$ (assuming $\alpha > 0$).

Now we look at the corresponding Fig. 9.-2. At the points A, B, and C, the absolute value of $\dot{q}$ is equal and q is zero. The higher spring-stiffness for $q > 0$ yields that the amplitude in the first half-period is smaller than in the second half-period and that Δt_1 is smaller than Δt_2. Therefore, the mean value of q is negative. We derive by our interpretation : $q_0 < 0$ if $\alpha > 0$. This is in accordance with eq. (5).

10. ELASTIC STABILITY (cf. sect. 3.4.4. of the lecture)

Introductory remarks

In the lecture we have seen that conjugate points are defined by $\delta^2 F = 0$. The interval between them is the smallest one for which $\delta^2 F$ vanishes. $\delta^2 F$ vanishes in any case if at least two neighbouring solutions intersect each other at the endpoints of the considered interval. But it is unknown for most of the problems whether such neighbouring solutions must always exist at conjugate points. This is, in the lecture, proved only for LAGRANGE-variations of one-dimensional problems without constraints, where the functional F depends on z and z' only. So we are sure that we can find out the conjugate points by looking for neighbouring solutions only in those cases where these conditions are satisfied.

On the other hand, when dealing with elastic stability, we are not originally interested in the question whether $\delta^2 F$ vanishes. This is only a necessary condition for instability. But we have to find out whether there is a position in which $U(q)$ is smaller (or at least equal) than at the considered, mostly trivial solution. Then, another solution must exist in the neighbourhood. This can be derived from results of later chapters.

Thus, for the problem of elastic stability, it is enough to ask for the existence of intersecting neighbouring solutions. As an example let us look at the four systems printed in Fig 10. -1

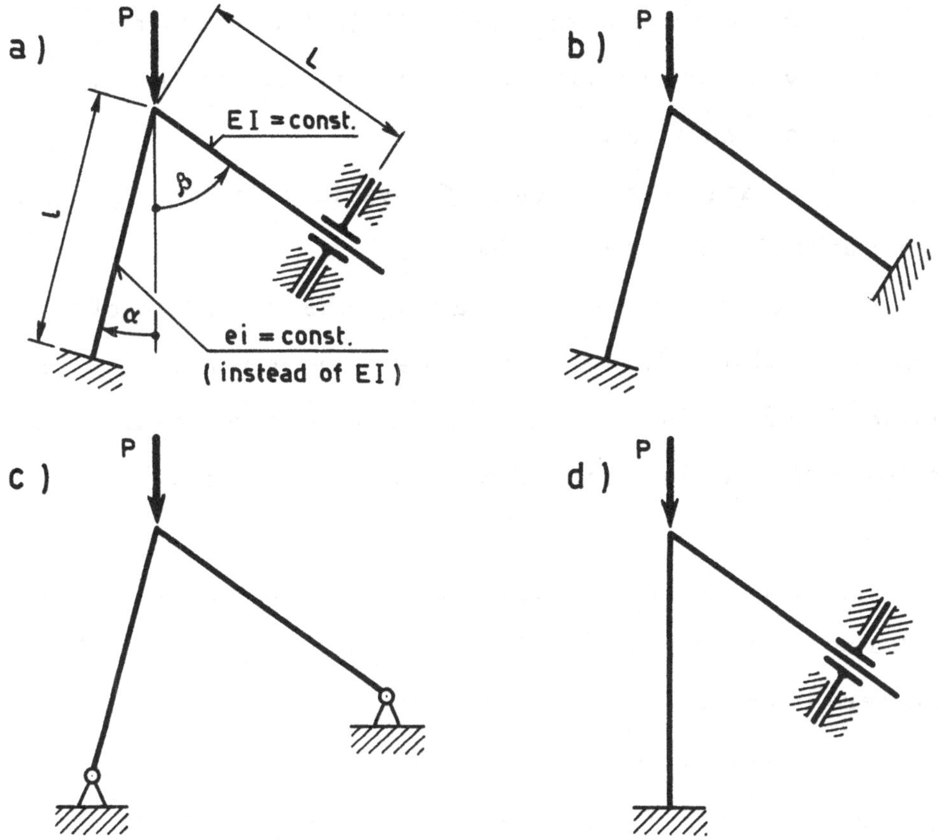

Fig. 10. - 1

The system a) is the only system for which we know that $\delta^2 F = 0$ is arrived first if neighbouring solutions intersect. But Problem a) is normally not a stability problem only in the specialized form d) it becomes a stability problem.

In case b) the following conditions are satisfied : The problem is one-dimensional (arc-length of the beams) the functional (potential energy) depends on z and z' only (z = angle of the inclination), and a LAGRANGE-variation of z is allowed. But we have to satisfy constraints because the endpoints cannot move .

In case c) the LAGRANGE-variation is lost, too. The tangent angle is free at the endpoints.

Hence, we state : In case d) the stability problem is connected with $\delta^2 F = 0$, and this is given if neighbouring solutions intersect in the endpoints. In case b) we can assume that this will be valid too, but without proof. In case c) this is not clear at all .

The technically most interesting problems of Fig. 10. -1 are the problems b) and c). Therefore, we will handle them here in detail.

Problem 10. -1 : Find out a formula for the determination of the critical load P_{crit} for the cases b) and c) of Fig. 10 -1 Assume that there is no tension and no shear-deformation in the beams.

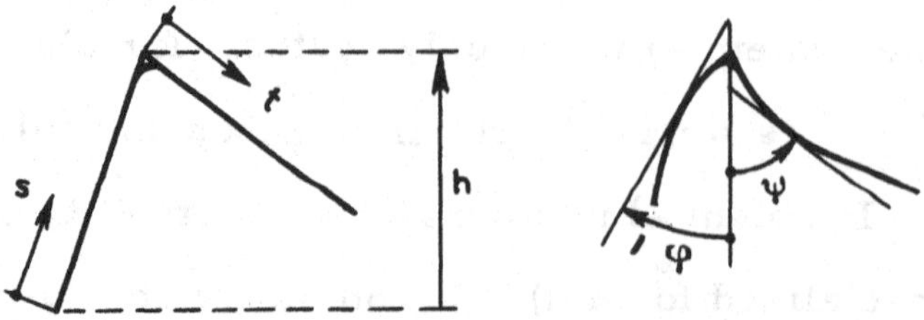

Fig. 10. -2

Fig. 10. - 2 shows the coordinates which we will use.

The solution has to satisfy three geometrical constraints:

$$\int_0^{\ell} \cos\varphi \, ds - \int_0^{L} \cos\psi \, dt - \ell\cos\alpha + L\cos\beta = 0 , \tag{1}$$

$$\int_0^{\ell} \sin\varphi \, ds + \int_0^{L} \sin\psi \, dt - \ell\sin\alpha - L\sin\beta = 0 , \tag{2}$$

$$\varphi(\ell) + \psi(0) = \alpha + \beta . \tag{3}$$

The last condition is nothing but a special smoothness-condition. Furthermore, we state that the potential energy tends to become a minimum :

$$U = Ph + \int_0^{\ell} \frac{ei}{2} (\varphi')^2 ds + \int_0^{L} \frac{EI}{2} (\psi')^2 dt \Longrightarrow \min , \tag{4}$$

where $\varphi' = \dfrac{d\varphi}{ds}$, $\psi' = \dfrac{d\psi}{dt}$. Because of $h = \int_0^{\ell} \cos\varphi \, ds$,

we get

$$U = \int_0^{\ell} \left[P\cos\varphi + \frac{ei}{2} (\varphi')^2 \right] ds + \int_0^{L} \frac{EI}{2} (\psi')^2 dt \Longrightarrow \min. \tag{5}$$

The minimum has to be reached under the side-conditions (1), (2) and (3). The constraints (1) and (2) are functionals of φ and ψ by themselves. We can treat them as to be functions in a space of an infinite number of coordinates. The same is possible with respect to U . Hence, we are allowed to apply the multiplier theory to (1) and (2) so that the corresponding variables λ are single values and no functions of s or t. Constraint (3) leads to a restriction for the variation :

$$\delta'\varphi(\ell) + \delta'\psi(0) = 0 . \tag{6}$$

So we get a restricted variation of a functional and of functions In case b) the variation is furthermore restricted by

$$\delta'\varphi(0) = \delta'\psi(L) = 0 . \tag{7}$$

Applying the multiplier-theory, the minimum property of the functional may be lost, but stationarity remains :

$$\left\{\begin{array}{l} F(\varphi,\psi,\lambda_1,\lambda_2) = \int_0^\ell f\,ds + \int_0^L g\,dt + h \Longrightarrow \text{stationary} \\ \text{where} \\ f \equiv f(\varphi,\varphi',\lambda_1,\lambda_2) = P\cos\varphi + \frac{ei}{2}(\varphi')^2 + \lambda_1\cos\varphi + \lambda_2\sin\varphi, \\ g \equiv g(\psi,\psi',\lambda_1,\lambda_2) = \frac{EI}{2}(\psi')^2 - \lambda_1\cos\psi + \lambda_2\sin\psi, \\ h \equiv h(\lambda_1,\lambda_2) = \\ = \lambda_1(L\cos\beta - \ell\cos\alpha) - \lambda_2(L\sin\beta + \ell\sin\alpha) \end{array}\right. \tag{8}$$

The angles are φ and ψ variable functions of or t resp. under the restrictions (6) and - possibly - (7), whereas the quantities λ_i can be varied as single values only .

Introducing

$$\varphi = \tilde{\varphi} + \varepsilon\bar{\varphi}\ , \quad \psi = \tilde{\psi} + \varepsilon\bar{\psi}\ , \quad \lambda_i = \tilde{\lambda}_i + \varepsilon\bar{\lambda}_i ,$$

where $\sim$ denotes the true solution and $-$ an arbitrary function or value restricted to

$$\bar{\varphi}(\ell) + \bar{\psi}(0) = 0 \quad \text{and in case b) by } \bar{\varphi}(0) = \bar{\psi}(L) = 0 , \tag{9}$$

the variation $\delta F = 0$ or $\dfrac{\partial F}{\partial \varepsilon} = 0$ yields

$$\int_0^\ell \left[\frac{\partial f}{\partial \varphi}\bigg|_{\tilde{\varphi},\tilde{\lambda}_i} \bar{\varphi} + \frac{\partial f}{\partial \varphi'}\bigg|_{\tilde{\varphi},\tilde{\lambda}_i} \bar{\varphi}' \right] ds +$$

$$+ \int_0^L \left[\frac{\partial g}{\partial \psi}\bigg|_{\tilde{\psi},\tilde{\lambda}_i} \bar{\psi} + \frac{\partial g}{\partial \psi'}\bigg|_{\tilde{\psi},\tilde{\lambda}_i} \bar{\psi}' \right] dt +$$

$$+ \sum_{i=1}^{2} \bar{\lambda}_i \cdot \left\{ \int_0^\ell \frac{\partial f}{\partial \lambda_i}\bigg|_{\tilde{\varphi},\tilde{\lambda}_j} ds + \int_0^L \frac{\partial g}{\partial \lambda_i}\bigg|_{\tilde{\psi},\tilde{\lambda}_j} dt + \frac{\partial h}{\partial \lambda_i}\bigg|_{\tilde{\lambda}_j} \right\} = 0 .$$

Partial integration leads to

$$\left[\frac{\partial f}{\partial \varphi'}\bigg|_{\tilde{\varphi},\tilde{\lambda}_i} \bar{\varphi} \right]_0^\ell + \left[\frac{\partial g}{\partial \psi'}\bigg|_{\tilde{\psi},\tilde{\lambda}_i} \bar{\psi} \right] +$$

$$+ \int_0^\ell \left[\frac{\partial f}{\partial \varphi} - \frac{d}{ds}\left(\frac{\partial f}{\partial \varphi'} \right) \right]\bigg|_{\tilde{\varphi},\tilde{\lambda}_i} \bar{\varphi}\, ds +$$

$$+ \int_0^L \left[\frac{\partial g}{\partial \psi} - \frac{d}{dt}\left(\frac{\partial g}{\partial \psi'} \right) \right]\bigg|_{\tilde{\psi},\tilde{\lambda}_i} \bar{\psi}\, dt + \bar{\lambda}_1 \left\{ \dots \right\} + \bar{\lambda}_2 \left\{ \dots \right\} = 0 ,$$

where the brackets $\{\dots\}$ are exactly identical with the original functional-constraints in our case, with the left sides of eqs (1) and (2) We remember that $\bar{\lambda}_i$ are arbitrary single values whereas $\bar{\varphi}$ and $\bar{\psi}$ are arbitrary functions .

For $\bar{\lambda}_1 \neq 0$, $\bar{\lambda}_2 = 0$, $\bar{\varphi} \equiv 0$, $\bar{\psi} \equiv 0$, we get eq (1), for $\bar{\lambda}_1 = 0$, $\bar{\lambda}_2 \neq 0$, $\bar{\varphi} \equiv 0$, $\bar{\psi} \equiv 0$, we get eq. (2), these equations have to be taken at $\varphi = \tilde{\varphi}$, $\psi = \tilde{\psi}$, $\lambda_i = \tilde{\lambda}_i$.

A LAGRANGE-variation of φ for $\bar{\psi} \equiv 0$, $\bar{\lambda}_i \equiv 0$ yields

$$\left[\frac{\partial f}{\partial \varphi} - \frac{d}{ds}\left(\frac{\partial f}{\partial \varphi'}\right)\right]\Bigg|_{\tilde{\varphi},\tilde{\lambda}_i} = 0 . \tag{10}$$

Correspondingly $\bar{\varphi} \neq 0$ leads to

$$\left[\frac{\partial g}{\partial \psi} - \frac{d}{dt}\left(\frac{\partial g}{\partial \psi'}\right)\right]\Bigg|_{\tilde{\psi},\tilde{\lambda}_i} = 0 . \tag{11}$$

At last we have to satisfy

$$- \frac{\partial f}{\partial \varphi'}\Bigg|_{\substack{s=0\\ \tilde{\varphi},\tilde{\lambda}_i}} \varphi(0) + \frac{\partial f}{\partial \varphi'}\Bigg|_{\substack{s=\ell\\ \tilde{\varphi},\tilde{\lambda}_i}} \varphi(\ell) - \frac{\partial g}{\partial \psi'}\Bigg|_{\substack{t=0\\ \tilde{\psi},\tilde{\lambda}_i}} \psi(0) +$$

$$+ \frac{\partial g}{\partial \psi'}\Bigg|_{\substack{t=L\\ \tilde{\psi},\tilde{\lambda}_i}} \psi(L) = 0 .$$

In case b) we have

$$\left.\begin{aligned} \varphi(0) &= \alpha\,, \\ \psi(L) &= \beta\,, \end{aligned}\right\} \text{(case b)} \qquad \begin{aligned}(12)\\(13)\end{aligned}$$

and $\bar{\varphi}(0) = \bar{\psi}(L) = 0$, so that the first and the last term vanish. At case c), $\bar{\varphi}(0)$ and $\bar{\psi}(L)$ are arbitrary values so that we derive

$$\left.\frac{\partial f}{\partial \varphi'}\right|_{\substack{s=0\\ \tilde{\varphi},\tilde{\lambda}_i}} = 0\ ; \quad \left.\frac{\partial g}{\partial \psi'}\right|_{\substack{t=L\\ \tilde{\psi},\tilde{\lambda}_i}} = 0 \quad \text{(case c)}. \qquad (14),\ (15)$$

Two terms remain Because of $\bar{\varphi}(\ell) = -\bar{\psi}(0)$ (arbitrary value) they yield

$$\left.\frac{\partial f}{\partial \varphi'}\right|_{\substack{s=\ell\\ \tilde{\varphi},\lambda_i}} + \left.\frac{\partial g}{\partial \psi'}\right|_{\substack{t=0\\ \tilde{\psi},\tilde{\lambda}_i}} = 0\,. \qquad (16)$$

Now we are prepared to insert f, g, and h according to eqs. (8) into eqs. (10), (11), (14), (15), and (16). Doing this, we replace $\tilde{\psi}$ by ψ, $\tilde{\varphi}$ by φ, $\tilde{\lambda}_i$ by λ_i. The above mentioned equations then read :

$$-P\sin\varphi - \lambda_1 \sin\varphi + \lambda_2 \cos\varphi - ei\,\varphi'' = 0\,, \qquad (17)$$

$$\lambda_1 \sin\psi + \lambda_2 \cos\psi - EI\,\psi'' = 0\,, \qquad (18)$$

$$\varphi'(0) = \psi'(L) = 0\,, \quad \text{(case c)} \qquad (19)$$

$$\text{(20)} \qquad ei\,\varphi'(\ell) + EI\,\psi'(0) = 0\,.$$

The equations (19), (20) can easily be interpreted : $\varphi'(0) = \psi'(L) = 0$ means : The bending-moment (proportional to φ' and ψ', resp) vanishes at the endpoints. This is true in case c) Equation (20) states that the bending-moment at $s = \ell$ is equal to the bending-moment at $t = 0$.

The following sets of equations describe the problem completely :

$$\left.\begin{array}{l}\text{case b)}\\ \text{case c)}\end{array}\right\}\ (1),(2),(3)\ \left\{\begin{array}{c}(12),(13)\\ (19)\end{array}\right\}\ (17)\ (18)\ (20)$$

The considered trivial solution which has to be checked whether it is stable is described by

$$\varphi = \alpha\,,\quad \psi = \beta\,,$$

which satisfies all equations if λ_i are chosen to be

$$\lambda_1 = k_1 = -\frac{\sin\alpha\,\cos\beta}{\sin(\alpha+\beta)}P\,;\quad \lambda_2 = k_2 = \frac{\sin\alpha\,\sin\beta}{\sin(\alpha+\beta)}P\,.$$

The next step is to look for disturbed solutions for

$$\varphi = \alpha + \delta\varphi\,,\quad \psi = \beta + \delta\psi\,,\quad \lambda_i = k_i + \delta\lambda_i\,,$$

where $\delta\varphi, \delta\psi, \delta\lambda_i$ are assumed to be very small so that we can linearize the equations

Eqs (1) and (2) read now

$$\int_0^{\ell}(\cos\alpha - \sin\alpha\ \delta\varphi)ds -$$

$$-\int_0^{L}(\cos\beta - \delta\psi\ \sin\beta)\,dt - \ell\cos\alpha + L\cos\beta = 0,$$

$$\int_0^{\ell}(\sin\alpha + \delta\varphi\cos\alpha)ds +$$

$$+\int_0^{L}(\sin\beta + \delta\psi\cos\beta)\,dt - \ell\sin\alpha - L\sin\beta = 0$$

or

$$\sin\alpha\int_0^{\ell}\delta\varphi\ ds - \sin\beta\int_0^{L}\delta\psi\ dt = 0\ ;$$

$$\cos\alpha\int_0^{\ell}\delta\varphi\ ds + \cos\beta\int_0^{L}\delta\psi\ dt = 0$$

or

$$(\sin\alpha\ \cos\beta + \sin\beta\ \cos\alpha)\int_0^{\ell}\delta\varphi\ ds = 0,$$

$$(\sin\alpha\ \cos\beta + \sin\beta\ \cos\alpha)\int_0^{L}\delta\psi\ dt = 0\,.$$

The brackets $\sin\alpha\ \cos\beta + \sin\beta\cos\alpha = \sin(\alpha+\beta)$ can be assumed to be different from zero : $\alpha+\beta \neq 0$, $\alpha+\beta \neq \pi$.

Then, we get :

(21), (22) $$\int_0^{\ell} \delta\varphi \, ds = 0 \; ; \quad \int_0^{L} \delta\psi \, dt = 0 .$$

Equation (3) gives us the well-known relation

(23) $$\delta\varphi(\ell) + \delta\psi(0) = 0 .$$

Equations (12), (13), (19) yield in

	case b),	case c)
(24)	$\delta\varphi(0) = 0$,	$\delta\varphi'(0) = 0$,
(25)	$\delta\psi(L) = 0$,	$\delta\psi'(L) = 0$.

Equation (17) can be expressed in the following manner

$$- P \sin\alpha - P\delta\varphi\cos\alpha - k_1\sin\alpha - k_1\delta\varphi\cos\alpha - \delta\lambda_1\sin\alpha +$$

$$+ k_2\cos\alpha - k_2\delta\varphi\sin\alpha + \delta\lambda_2\cos\alpha - ei\,\delta\varphi'' = 0,$$

where

$$- P\sin\alpha - k_1\sin\alpha + k_2\cos\alpha =$$

$$= P\left\{ -\sin\alpha + \frac{\sin^2\alpha\cos\beta}{\sin(\alpha+\beta)} + \frac{\sin\alpha\cos\alpha\sin\beta}{\sin(\alpha+\beta)} \right\} =$$

$$= P\left\{ -\sin\alpha + \sin\alpha \right\} = 0$$

$$P\cos\alpha + k_1\cos\alpha + k_2\cos\alpha =$$

$$= P\left\{+\cos\alpha - \frac{\sin\alpha\cos\alpha\cos\beta - \sin^2\alpha\sin\beta}{\sin(\alpha+\beta)}\right\} =$$

$$= P\,\frac{\sin\beta}{\sin(\alpha+\beta)}\ .$$

This leads to

$$ei\,\delta\varphi'' + P\,\frac{\sin\beta}{\sin(\alpha+\beta)}\,\delta\varphi = -\delta\lambda_1\sin\alpha + \delta\lambda_2\cos\alpha. \quad (26)$$

Similarly eq (18) yields

$$EI\,\delta\psi'' + P\,\frac{\sin\alpha}{\sin(\alpha+\beta)}\,\delta\psi = \delta\lambda_1\sin\beta + \delta\lambda_2\cos\beta. \quad (27)$$

At last, eq (20) leads to

$$ei\,\delta\varphi'(\ell) + EI\,\delta\psi'(0) = 0\ . \quad (28)$$

This set of eqs (21) to (28) must be satisfied by any neighbouring solution.

When solving this problem we remember that $\delta\varphi$ and $\delta\psi$ are functions of s or t, resp., whereas $\delta\lambda_i$ is a single value which is of no interest. Hence eqs (26) (27) are equations of a vibration with the solution

$$\delta\varphi = \delta a + \delta A \sin\left(\sqrt{P\gamma}\ \frac{s}{\ell}\right) + \delta B \cos\left(\sqrt{P\gamma}\ \frac{s}{\ell}\right)$$

and

$$\delta\psi = \delta b + \delta C \sin\left(\sqrt{P\varepsilon}\ \frac{L-t}{L}\right) + \delta D \cos\left(\sqrt{P\varepsilon}\ \frac{L-t}{L}\right),$$

where δa, δb, $\delta A, \ldots \delta D$ are arbitrary values, whereas γ and ε are parameters of the system

$$\gamma = \frac{\ell^2 \sin\beta}{ei\ \sin(\alpha+\beta)}\ ; \quad \varepsilon = \frac{L^2 \sin\alpha}{EI\ \sin(\alpha+\beta)}\ . \tag{29}$$

From eqs (24), (25) we get in case b) :

$$\delta a = -\delta B\ ; \quad \delta b = -\delta D$$

and in case c)

$$\delta A = \delta C = 0\ .$$

The next step is to satisfy eqs (21), (22) This yields for case b)

$$-\delta B \cdot \ell + \delta A \frac{\ell}{\sqrt{P\gamma}}\left(1-\cos\sqrt{P\gamma}\right) + \delta B \frac{\ell}{\sqrt{P\gamma}} \sin\sqrt{P\gamma} = 0,$$

$$-\delta D \cdot L + \delta C \frac{L}{\sqrt{P\varepsilon}}\left(1-\cos\sqrt{P\varepsilon}\right) + \delta D \frac{L}{\sqrt{P\varepsilon}} \sin\sqrt{P\varepsilon} = 0.$$

Hence,

$$\delta A = \frac{\sqrt{P\gamma}}{1-\cos\sqrt{P\gamma}}\left(1-\frac{\sin\sqrt{P\gamma}}{\sqrt{P\gamma}}\right)\delta B,$$

$$\delta C = \frac{\sqrt{P\varepsilon}}{1-\cos\sqrt{P\varepsilon}}\left(1-\frac{\sin\sqrt{P\varepsilon}}{\sqrt{P\varepsilon}}\right)\delta D$$

or

$$\delta\varphi = \delta B\cdot\Phi \;;\; \delta\psi = \delta D\cdot\Psi\,, \tag{30}$$

where

$$\left.\begin{aligned}\Phi &= -1+\cos\left(\sqrt{P\gamma}\,\frac{s}{\ell}\right)+\frac{\sqrt{P\gamma}-\sin\sqrt{P\gamma}}{1-\cos\sqrt{P\gamma}}\sin\left(\sqrt{P\gamma}\,\frac{s}{\ell}\right),\\ \Psi &= -1+\cos\left(\sqrt{P\gamma}\,\frac{L-t}{L}\right)+\frac{\sqrt{P\varepsilon}-\sin\sqrt{P\varepsilon}}{1-\cos\sqrt{P\varepsilon}}\sin\left(\sqrt{P\varepsilon}\,\frac{L-t}{L}\right).\end{aligned}\right\}\text{(case b)}$$

Similarly, we derive from case c)

$$\delta a\,\ell+\frac{\delta B\ell}{\sqrt{P\gamma}}\sin\sqrt{P\gamma}=0\,,\qquad \delta b\,L+\frac{\delta DL}{\sqrt{P\varepsilon}}\sin\sqrt{P\varepsilon}=0\,,$$

hence, eq. (30) where

$$\Phi = -\frac{\sin\sqrt{P\gamma}}{\sqrt{P\gamma}} + \cos\left(\sqrt{P\gamma}\,\frac{s}{\ell}\right),$$

$$\Psi = -\frac{\sin\sqrt{P\varepsilon}}{\sqrt{P\varepsilon}} + \cos\left(\sqrt{P\varepsilon}\,\frac{L-t}{L}\right).$$

As we see, these functions Φ and Ψ are given functions $\Phi = \Phi(s, P)$ and $\Psi = \Psi(t,P)$. They satisfy with eq. (30) nearly the complete set of equations (21) to (28) except eqs (23) and (28) These last relations read now :

$$\delta B \cdot \Phi(\ell, P) + \delta D \cdot \Psi(0,P) = 0,$$

$$\delta B\, ei\, \frac{\partial\Phi}{\partial s}\bigg|_{\ell,P} + \delta D \cdot EI\, \frac{\partial\Psi}{\partial t}\bigg|_{0,P} = 0.$$

They form an homogeneous system of linear equations Thus, $\delta B = \delta D = 0$ is a trivial solution. But this result is of no interest We look for solutions in which δB and δD do not vanish. Hence, the determinant of the coefficients which in this situation, depend on P only has to vanish. The first value of P at which this is reached, is our critical load. The equation for P_{crit} which can be derived by this way, reads

$$EI\,\Phi(\ell, P_{crit})\,\frac{\partial\Psi}{\partial t}\bigg|_{0,P_{crit}} - ei\,\Psi(0, P_{crit})\,\frac{\partial\Phi}{\partial s}\bigg|_{\ell,P_{crit}} = 0\,.$$

In case c) (which is simpler) this relation leads to ($P \equiv P_{crit}$)

$$EI \cdot \left[-\frac{\sin\sqrt{P\gamma}}{\sqrt{P\gamma}} + \cos\sqrt{P\gamma}\right] \cdot \frac{\cos\sqrt{P\varepsilon}}{L} -$$

$$-\,ei\left[-\frac{\sin\sqrt{P\varepsilon}}{\sqrt{P\varepsilon}} + \cos\sqrt{P\varepsilon}\right] \cdot \left(-\frac{\cos\sqrt{P\varepsilon}}{\ell}\right) = 0$$

or multiplying by $\ell \cdot L \cdot \sqrt{P\gamma} \cdot \sqrt{P\varepsilon}$:

$$EI\,\ell\left(\sqrt{P\gamma}\,\cos\sqrt{P\gamma} - \sin\sqrt{P\gamma}\right)\cos\sqrt{P\varepsilon}\; +$$

$$+\;ei\,L\left(\sqrt{P\varepsilon}\,\cos\sqrt{P\varepsilon} - \sin\sqrt{P\varepsilon}\right)\cos\sqrt{P\gamma} = 0.$$

This relation is satisfied for $P=0$. But this can be excluded, looking at eqs. (26), (27), (21), (22) we see that this would lead to $\delta\varphi = \delta\psi = 0$, hence, to the trivial solution. Thus, we have the look for a solution with $P>0$. This can be done numerically. In the case of $\gamma = \varepsilon$, $EI\ell = k\,ei\,L$ which includes

$$\frac{L \sin\alpha}{\ell \sin\beta} = k ,$$

we get with $\sqrt{P\gamma} = \sqrt{P\varepsilon} = k$

$$k(\lambda \cos\lambda - \sin\lambda)\cos\lambda + (\lambda \cos\lambda - \sin\lambda)\cos\lambda = 0 \quad \text{or}$$

$$(k+1)(\lambda \cos\lambda - \sin\lambda)\cos\lambda = 0,$$

where we are interested in

$$\min(\lambda), \quad \lambda > 0$$

From $\cos\lambda = 0$ we get $\lambda = \pi/2$, whereas $\lambda\cos\lambda - \sin\lambda = 0$ yields no smaller solution : $\lambda = \tan\lambda$ needs $\lambda > \pi/2$. Hence, $\lambda = \pi/2$ is the first critical value of λ, hence :

$$P_{crit} = \frac{\pi^2}{4\varepsilon} .$$

The physical interpretation of this result is quite trivial : The geometry of the system is given in such a way that both beams would have reached the critical point under the same value of P, if the system would contain a hinge at $s = \ell / t = 0$ Hence, the buckling is not hindered.

11. CANONICAL EQUATIONS (cf. sect. 3. 5 of the lecture)

In this chapter we will calculate the canonical equations which belong to a simple problem. As we will see from the result, they will not enable us to find out the equations of motion in a simpler way than, for example by using LAGRANGE-equations of 2nd kind. But the main reason for establishing the canonical equations is the fact that they lead to JACOBI's integration theory which does not belong to our material

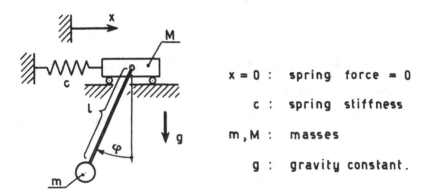

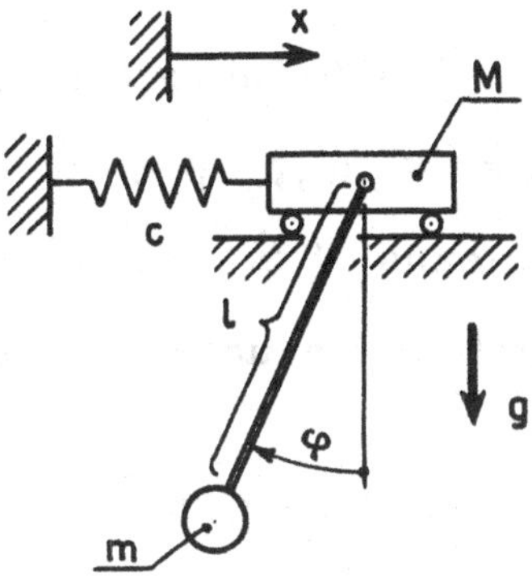

$x = 0$: spring force = 0

c : spring stiffness

m, M : masses

g : gravity constant.

Fig 11. -1

<u>Problem 11. -1</u> : Find out the canonical equations describing the problem printed in Fig. 11.-1.

x and φ are free coordinates. The kinetic potential L expresses by $x, \varphi, \dot{x}, \dot{\varphi}$ is (cf problem 6. 2-1) :

$$L = T - U = \underbrace{\frac{1}{2}(M+m)\dot{x}^2 - m\ell\dot{x}\dot{\varphi}\cos\varphi + \frac{1}{2}m\ell^2\dot{\varphi}^2}_{T} -$$

$$\underbrace{-\frac{1}{2}cx^2 + mg\ell\cos\varphi}_{-U} ;$$

The special application of the LEGENDRE-transformation derived in the lecture, states that the impulses p_i are

$$(1) \qquad p_i = \frac{\partial L}{\partial \dot{q}_i} \quad \text{or} \quad \begin{cases} p_\varphi = -m\ell\dot{x}\cos\varphi + m\ell^2\dot{\varphi}, \\ p_x = (M+m)\dot{x} - m\ell\dot{\varphi}\cos\varphi . \end{cases}$$

The HAMILTON's function $H = T + U$ has to be expressed in terms of q_i and p_i, that means : by $x, \varphi, p_x, p_\varphi$. In L, the kinetic energy T is expressed by means of $\dot{x}$ and $\dot{\varphi}$. Therefore, we need the inverse transformational equations to eqs (1):

$$(2) \qquad \begin{aligned} \dot{x} &= \left[p_x + p_\varphi \frac{\cos\varphi}{\ell} \right] \cdot (M + m\sin^2\varphi)^{-1}, \\ \dot{\varphi} &= \left[p_x \frac{\cos\varphi}{\ell} + p_\varphi \frac{M+m}{m\ell^2} \right] \cdot (M + m\sin^2\varphi)^{-1} . \end{aligned}$$

With these relations, we can write

$$H = T + U = \frac{1}{2}(M+m)\dot{x}^2 - m\ell\dot{x}\dot{\varphi}\cos\varphi +$$

$$+ \frac{1}{2}m\ell^2\dot{\varphi}^2 + \frac{1}{2}cx^2 - mg\ell\cos\varphi =$$

$$= \frac{1}{2}\left[p_x^2 + \frac{2}{\ell} p_x p_\varphi \cos\varphi + p_\varphi^2 \frac{M+m}{m\ell^2}\right]\left(M + m\sin^2\varphi\right)^{-1} + \\ + \frac{1}{2} c x^2 - mg\ell\cos\varphi . \tag{3}$$

The canonical equations are

$$\dot{q}_x = \dot{x} = \frac{\partial H}{\partial p_x} \; ; \; \dot{q}_\varphi = \dot{\varphi} = \frac{\partial H}{\partial p_\varphi} \; , \; \text{and} \qquad \dot{p}_i = - \frac{\partial H}{\partial q_i} . \quad (4), (5)$$

Normally, we would have to stop here. But we want to check whether eqs. (4) and (5) yield the normal result. Deriving with respect to p_x and p_φ, we easily see that eqs. (4) lead to eqs. (2). Using eqs. (1) and (3), the relation $\dot{p}_x = - \frac{\partial H}{\partial x}$ yields

$$(M+m)\ddot{x} - m\ell\ddot{\varphi}\cos\varphi + m\ell\dot{\varphi}^2\sin\varphi = - cx .$$

This is the horizontal equilibrium of forces. The relation $\dot{p}_\varphi = - \frac{\partial H}{\partial \varphi}$ leads to

$$-m\ell\ddot{x}\cos\varphi + m\ell\dot{x}\dot{\varphi}\sin\varphi + m\ell^2\ddot{\varphi} = \frac{1}{\ell} p_x p_\varphi \sin\varphi\,(M + m\sin^2\varphi)^{-1} + \\ + \left[p_x^2 + \frac{2}{\ell} p_x p_\varphi \cos\varphi + p_\varphi^2 \frac{M+m}{m\ell^2}\right](M + m\sin^2\varphi)^{-2} - mg\ell\sin\varphi .$$

After some lengthy calculations and substituting for p_x and p_φ the right sides of eqs. (1) this relation yields

$$\ddot{x}\cos\varphi + \ell\ddot{\varphi} + g\sin\varphi = 0,$$

which, multiplied by $m\ell$, is the equilibrium of torques around the link at the mass M.

12. ELASTOMECHANICS (cf. sect. 4.1 of the lecture)

12.1. Plate theory

We want to deliver a plate theory starting from the basic inequality

(1) $$U(q) - \Lambda_{ex}(\hat{F} < x >) \geq U(\hat{q}) - \Lambda_{ex}(\hat{F} < \hat{x} >).$$

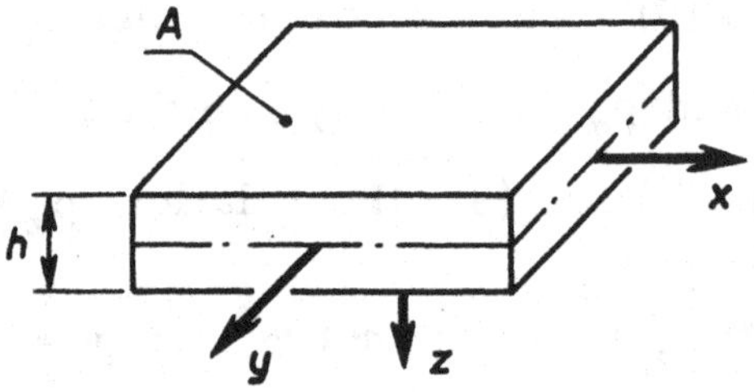

Fig. 12.1-1

Fig. 12.1-1 shows the coordinates. The plate theory derived here starts from the following assumptions :

1. The middle surface of the plate which is identical with the the x,y-plane in the original unloaded state is not loaded in x- and y - directions. That means :

   $$\sigma_{xx} = \sigma_{yy} = \sigma_{xy} = 0 \quad \text{at} \quad z = 0 .$$

2. σ_{zz} vanishes identically : $\sigma_{zz} = 0$.

3. Straight lines normal to the original middle plane remain to be straight lines normal to the (deformed) middle surface.

4 Displacements of the middle surface in x- or y-directions do not exist (cf. assumption 1) whereas a displacement in z-direction exists and will be called w.

5 A load q per unit area in z-direction may act upon the middle surface.

6. The derived formulae have to be linear in w, that prescribes that the functional itself can only be quadratic in w.

The theory is completely prescribed by eq. (1) and by the six assumptions.

First, we have to look at the geometrical conditions The unit vector normal to the middle surface $w(x,y)$ is given by :

$$\underline{n} = \frac{1}{\sqrt{1+\left(\frac{\partial w}{\partial x}\right)^2+\left(\frac{\partial w}{\partial y}\right)^2}} \begin{bmatrix} -\frac{\partial w}{\partial x} \\ -\frac{\partial w}{\partial y} \\ 1 \end{bmatrix} \approx \begin{bmatrix} -\frac{\partial w}{\partial x} \\ -\frac{\partial w}{\partial y} \\ 1 \end{bmatrix} .$$

A point which is originally situated at $\underline{r}^0 = \begin{bmatrix} x \\ y \\ z \end{bmatrix}$, will be displaced to the point

(2) $$\underline{r} = \begin{bmatrix} x \\ y \\ w \end{bmatrix} + \underline{n} \cdot z\,(\bar{\varepsilon}_{zz} + 1)\,,$$

where $\bar{\varepsilon}_{zz}$ is the average strain ε_{zz} between $z = 0$ and z.

(3) $$\bar{\varepsilon}_{zz} = \frac{1}{z} \int_0^z \varepsilon_{zz} \Big|_{x,y=\mathrm{const.}} dz\,.$$

From eq. (2) we derive

$$\underline{r} = \begin{bmatrix} x - \dfrac{\partial w}{\partial x}\, z\,(\bar{\varepsilon}_{zz} + 1) \\ y - \dfrac{\partial w}{\partial y}\, z\,(\bar{\varepsilon}_{zz} + 1) \\ w + z\,(\bar{\varepsilon}_{zz} + 1) \end{bmatrix}$$

or for the vector $\underline{u}$ of the displacements

(4) $$\underline{u} = \begin{bmatrix} -z\,(1 + \bar{\varepsilon}_{zz})\, \dfrac{\partial w}{\partial x} \\ -z\,(1 + \bar{\varepsilon}_{zz})\, \dfrac{\partial w}{\partial y} \\ w + z\,\bar{\varepsilon}_{zz} \end{bmatrix}.$$

In a linear theory, we are allowed to start from

(5) $$\varepsilon_{ij} = \frac{1}{2}\left(\frac{\partial u_i}{\partial x_j} + \frac{\partial u_j}{\partial x_i}\right).$$

This formula yields for example :

$$\varepsilon_{xx} = -z \frac{\partial \bar{\varepsilon}_{zz}}{\partial x} \frac{\partial w}{\partial x} - z\left(1+\bar{\varepsilon}_{zz}\right)\frac{\partial^2 w}{\partial x^2} \approx -z \frac{\partial^2 w}{\partial x^2} .$$

In this expression we can drop the terms containing $\bar{\varepsilon}_{zz}$, because $\bar{\varepsilon}_{zz}$ will be an at least linear function of w , so that these terms containing $\bar{\varepsilon}_{zz}$ would lead to nothing but quadratic terms of w. Each ε_{ij} will, as we will see later, contain only linear, quadratic , cubic, etc. terms of w (no constant term). When writing down $U(q)$, each of these ε_{ij} will be multiplied by another ε_{ij}, so that $U(q)$ is quadratic in w if ε_{ij} is linear in w . Looking at our last assumption, we recognize that, for this reason, quadratic terms in ε_{ij} can be dropped. So we are allowed to use

$$\varepsilon_{xx} = -z \frac{\partial^2 w}{\partial x^2} \quad ; \quad \varepsilon_{yy} = -z \frac{\partial^2 w}{\partial y^2} . \tag{6}$$

The same argument enabled us to linearize $\underline{n}$ (cf. between eqs. (1) and (2)) and to deal with eq. (5) .

Assumption 2 gives us the possibility to express ε_{zz} by means of ε_{xx} and ε_{yy} :

$$E\,\varepsilon_{xx} = \sigma_{xx} - \nu\,\sigma_{yy} , \tag{7}$$

$$E\,\varepsilon_{yy} = \sigma_{yy} - \nu\,\sigma_{xx} , \tag{8}$$

(9) $$E\varepsilon_{zz} = -\nu(\sigma_{xx}+\sigma_{yy})$$

Eqs. (7) (8) yield $E(\varepsilon_{xx}+\varepsilon_{yy}) = (1-\nu)(\sigma_{xx}+\sigma_{yy})$ or

$$E(\varepsilon_{xx}+\varepsilon_{yy}) = -\frac{1-\nu}{\nu}\, E\varepsilon_{zz} \quad \text{or}$$

(10) $$\varepsilon_{zz} = -\frac{\nu}{1-\nu}(\varepsilon_{xx}+\varepsilon_{yy}) = +\frac{\nu}{1-\nu}\,\Delta w\cdot z\,.$$

Equation (3) leads to

(11) $$\bar{\varepsilon}_{zz} = \frac{\nu}{1-\nu}\,\frac{\Delta w}{2}\cdot z$$

so that we have instead of eq (4) :

$$\underline{u} = \begin{bmatrix} -z\left(1+\cancel{\frac{\nu}{1-\nu}\frac{\Delta w}{2}z}\right)\frac{\partial w}{\partial x} \\ -z\left(1+\cancel{\frac{\nu}{1-\nu}\frac{\Delta w}{2}z}\right)\frac{\partial w}{\partial y} \\ w+\frac{\nu}{1-\nu}\frac{\Delta w}{2}z^2 \end{bmatrix}$$

$\nearrow$: because of quadratic dependence on w

or

(12) $$\underline{u} \approx \begin{bmatrix} -z\,\frac{\partial w}{\partial x} \\ -z\,\frac{\partial w}{\partial y} \\ w+\frac{\nu}{1-\nu}\,\frac{\Delta w}{2}\,z^2 \end{bmatrix}.$$

By eq. (5) we derive from eq. (12) that the matrix of the deformations ε_{ij} is equal to

$$[\varepsilon_{ij}] = \begin{bmatrix} -z\dfrac{\partial^2 w}{\partial x^2} & -z\dfrac{\partial^2 w}{\partial x\,\partial y} & \dfrac{z^2}{4}\dfrac{\nu}{1-\nu}\dfrac{\partial \Delta w}{\partial x} \\ -z\dfrac{\partial^2 w}{\partial x\,\partial y} & -z\dfrac{\partial^2 w}{\partial y^2} & \dfrac{z^2}{4}\dfrac{\nu}{1-\nu}\dfrac{\partial \Delta w}{\partial y} \\ \dfrac{z^2}{4}\dfrac{\nu}{1-\nu}\dfrac{\partial \Delta w}{\partial x} & \dfrac{z^2}{4}\dfrac{\nu}{1-\nu}\dfrac{\partial \Delta w}{\partial y} & z\dfrac{\nu}{1-\nu}\Delta w \end{bmatrix}. \tag{13}$$

The potential energy U can be written as follows (cf. sect. 6. 2):

$$U(q) = \frac{1}{2}\iiint^{V} \sigma_{ij}\,\varepsilon_{ij}\,dV = \text{(summation)}$$

$$= \frac{1}{2}\iiint^{V}\Big\{\frac{E}{1-\nu^2}\Big[\varepsilon_{xx}(\varepsilon_{xx}+\nu\varepsilon_{yy}) + \varepsilon_{yy}(\varepsilon_{yy}+\nu\varepsilon_{xx})\Big] +$$

$$+ \frac{2E}{1+\varepsilon}\left(\varepsilon_{xy}^2 + \varepsilon_{yz}^2 + \varepsilon_{zx}^2\right)\Big\}\,dV.$$

Here we have used : $\sigma_{ij} = 2G\varepsilon_{ij} = \dfrac{E}{1+\nu}\varepsilon_{ij}$ if $i \neq j$ and

$$\sigma_{zz} = 0\,,\quad \sigma_{xx} = \frac{E}{1-\nu^2}(\varepsilon_{xx}+\nu\varepsilon_{yy})\,;\quad \sigma_{yy} = \frac{E}{1-\nu^2}(\varepsilon_{yy}+\nu\varepsilon_{xx})\,.$$

This can be seen from eqs. (7) and (8). At this place we notice that indeed each ε_{ij} is multiplied by another ε_{ij}, what was mentioned above. Rearranging the integrand and substituting for ε_{ij} the terms of eq. (13) we reach :

$$U(q) = \frac{1}{2}\iiint^{V} \frac{E}{1-\nu^2}\left\{\varepsilon_{xx}^2 + \varepsilon_{yy}^2 + 2\varepsilon_{xy}^2 + 2\varepsilon_{yz}^2 + 2\varepsilon_{xz}^2 + \right.$$

$$\left. + \nu\left(2\varepsilon_{xx}\varepsilon_{yy} - 2\varepsilon_{xy}^2 - 2\varepsilon_{yz}^2 - 2\varepsilon_{xz}^2\right)\right\} dV =$$

$$= \frac{1}{2}\iint^{A}\left\{\int_{-\frac{h}{2}}^{+\frac{h}{2}} \frac{E}{1-\nu^2}\left[z^2\left(\frac{\partial^2 w}{\partial x^2}\right)^2 + z^2\left(\frac{\partial^2 w}{\partial y^2}\right)^2 + 2z^2\left(\frac{\partial^2 w}{\partial x \partial y}\right)^2 + \right.\right.$$

$$+ \frac{z^4}{8}\,\frac{\nu^2}{1-\nu}\left(\left(\frac{\partial \Delta w}{\partial x}\right)^2 + \left(\frac{\partial \Delta w}{\partial y}\right)^2\right) +$$

$$\left.\left. + 2\nu z^2\left(\frac{\partial^2 w}{\partial x^2}\,\frac{\partial^2 w}{\partial y^2} - \left(\frac{\partial^2 w}{\partial x \partial y}\right)^2\right)\right] dz\right\} dA =$$

$$= \frac{1}{2}\iint^{A} \frac{E}{1-\nu^2}\left\{\frac{h^3}{12}\left[\left(\frac{\partial^2 w}{\partial x^2}\right)^2 + \left(\frac{\partial^2 w}{\partial y^2}\right)^2 + 2\nu\left(\frac{\partial^2 w}{\partial x \partial y}\right) + \right.\right.$$

$$\left.\left. + 2\nu\left(\frac{\partial^2 w}{\partial x^2}\,\frac{\partial^2 w}{\partial y^2} - \left(\frac{\partial^2 w}{\partial x \partial y}\right)^2\right)\right] + \frac{h^5\nu^2}{640(1-\nu)}\left[\left(\frac{\partial \Delta w}{\partial x}\right)^2 + \left(\frac{\partial \Delta w}{\partial y}\right)^2\right]\right\} dA.$$

For further applications we use the notation : $w_{,\alpha} \equiv \dfrac{\partial w}{\partial x_\alpha}$

and the summation convention. Then we get for $\alpha = x, y$:

$$U(q) = \frac{1}{2} \iint^{A} \frac{E}{1-\nu^2} \left\{ \frac{h^3}{12} \left[w_{,\alpha\beta}\, w_{,\alpha\beta} + \nu\, \epsilon_{3\alpha\beta}\, \epsilon_{3\gamma\delta}\, w_{,\alpha\gamma} w_{,\beta\delta} + \right.\right.$$

$$\left.\left. + \frac{h^5 \nu^2}{640(1-\nu)}\, w_{,\alpha\alpha\gamma}\, w_{,\beta\beta\gamma} \right] \right\} dA .$$

This is the first part of our basic equation (1). The next term is $\Lambda_{ex}\left(\hat{F} < x >\right)$, this expression is composed of the following parts :

1 the load $\hat{q}$ acting in the z-direction, parallel to w, so that its "potential energy" is $-\iint^{A} \hat{q}\, w\, dA$,

2. a shear force $\hat{P}$ per unit arc length at the boundary acting parallel to w, which yields $-\oint^{B} \hat{P} w\, ds$ (s = arc-length of the boundary),

3. a moment $\underline{\hat{M}} = \hat{M}_x \underline{e}_x + \hat{M}_y \underline{e}_y = M_\alpha \underline{e}_\alpha$ acting at the boundary. It works against the rotation of the middle surface which is given by $\frac{1}{2}\operatorname{rot} \underline{u}$ taken at the middle surface :

$$\operatorname{rot} \underline{u} = \nabla \times \underline{u} = \underline{e}_a \times \frac{\partial}{\partial x_a} u_b \underline{e}_b = \epsilon_{abc} \frac{\partial u_b}{\partial x_a} \underline{e}_c = \epsilon_{abc} u_{b,a} \underline{e}_c .$$

The potential energy of the moment $\hat{M}$ then is

$$-\underline{\hat{M}} \cdot \frac{1}{2} \operatorname{rot} \underline{u} = -\frac{1}{2} \hat{M}_\alpha\, \underline{e}_\alpha \cdot \epsilon_{abc}\, u_{b,a}\, \underline{e}_c =$$

$$= -\frac{1}{2}\hat{M}_\alpha \epsilon_{ab\alpha} u_{b,a} =$$ (a <u>or</u> b must be 3, the other is then not equal to 3)

$$= -\frac{1}{2}\hat{M}_\alpha \epsilon_{3\beta\alpha} u_{\beta,3} - \hat{M}_\alpha \epsilon_{\beta 3\alpha} u_{3,\beta} =$$ (given by eq. (12))

$$= -\frac{1}{2}\hat{M}_\alpha \epsilon_{3\beta\alpha}(-w_{,\beta}) - \frac{1}{2}\hat{M}_\alpha \epsilon_{\beta 3\alpha} w_{,\beta} =$$

$$= \hat{M}_\alpha w_{,\beta} \cdot \frac{1}{2}\left(\epsilon_{3\beta\alpha} - \epsilon_{\beta 3\alpha}\right) =$$

$$= \hat{M}_\alpha w_{,\beta} \epsilon_{3\beta\alpha} = -\epsilon_{3\alpha\beta} \hat{M}_\alpha w_{,\beta} .$$

Now we have expressed all terms of the left side of eq. (1). This left side has to tend to a minimum (for given values of $\hat{q}$, $\hat{P}$, $\hat{M}_\alpha$), if we want to get a very good approximation of the true solution which is normally not included by our assumptions.

Thus, we state with :

(15) $$D = \frac{E h^3}{12(1-\nu^2)} :$$

$$F = \frac{1}{2}\iint^A D\left\{ w_{,\alpha\beta} w_{,\alpha\beta} + \nu \epsilon_{3\alpha\beta} \epsilon_{3\gamma\delta} w_{,\alpha\gamma} w_{,\beta\delta} + \right.$$

$$\left. + \frac{3}{160} h^2 \frac{\nu^2}{1-\nu} w_{,\alpha\alpha\gamma} w_{,\beta\beta\gamma} \right\} dA - \iint^A \hat{q}\, w\, dA -$$

$$- \oint^B \left[\hat{P} w + \epsilon_{3\alpha\beta} \hat{M}_\alpha w_{,\beta} \right] ds \Longrightarrow \min$$

We will calculate the EULER-LAGRANGE equations for this minimum principle for the case that $D = \text{const}$. Further on we assume that it is allowed to drop the term $\frac{3}{160} h^2 \frac{\nu^2}{1-\nu} w_{,\alpha\alpha\gamma} w_{,\beta\beta\gamma}$. This is a usual assumption but it does not follow from our six assumptions. It is reasonable if the wave-length of any part of the solution is much larger than h. This does not depend on the values of w.

Under these assumptions, the variation of F leads to a "best approximation" of the true result if we start from $w + \varepsilon \bar{w}$ instead of w and use $\partial F / \partial \varepsilon = 0$. This yields

$$\iint^{A} \left\{ D \left[w_{,\alpha\beta} \bar{w}_{,\alpha\beta} + \nu \left(w_{,\alpha\alpha} \bar{w}_{,\beta\beta} - w_{,\alpha\beta} \bar{w}_{,\alpha\beta} \right) \right] - \hat{q} \bar{w} \right\} dA -$$

$$- \oint^{B} \left(\hat{P} \bar{w} + \epsilon_{3\alpha\beta} \hat{M}_\alpha \bar{w}_{,\beta} \right) ds = 0 .$$

Using twice the GAUSS-GREEN-theorem we get with $\underline{\nu}$ = normal unit vector at B, directed outward:

$$\iint^{A} \left[D w_{,\alpha\alpha\beta\beta} - \hat{q} \right] \bar{w} \, dA + \oint^{B} \left\{ D \left[w_{,\alpha\beta} \bar{w}_{,\alpha} \nu_\beta + \nu \left(w_{,\alpha\alpha} \bar{w}_{,\beta} \nu_\beta - \right.\right.\right.$$

$$\left.\left.\left. - w_{,\alpha\beta} \bar{w}_{,\beta} \nu_\alpha \right) - w_{,\alpha\beta\beta} \bar{w} \nu_\alpha \right] - \hat{P} \bar{w} - \epsilon_{3\alpha\beta} \hat{M}_\alpha \bar{w}_{,\beta} \right\} ds = 0 . \tag{16}$$

A LAGRANGE-variation in the area A yields $D\, w_{,\alpha\alpha\beta\beta} - q = 0$, which is better known as

$$\frac{E h^3}{12(1-\nu^2)} \Delta\Delta w - \hat{q} = 0 \quad (D = \text{const.}) \tag{17}$$

This is a first result Boundary conditions for not-supported boundaries are missing up to now. If we want to derive them we must start from

$$\bar{w}_{,\nu} = \bar{w}_{,\alpha} \nu_\alpha \ ; \quad \bar{w}_{,t} = \bar{w}_{,\alpha} t_\alpha , \quad \text{and}$$
$$\bar{w}_{,\alpha} = \bar{w}_{,\nu} \nu_\alpha + \bar{w}_{,t} t_\alpha . \tag{18}$$

The vectors $\underline{\nu}$ and $\underline{t}$ are unit vectors in the x, y-plane normal and tangential to the boundary. They are connected by

$$\nu_\alpha t_\alpha = 0 , \quad \nu_\alpha = \epsilon_{3\alpha\beta} t_\beta , \quad t_\alpha = -\epsilon_{3\alpha\beta} \nu_\beta . \tag{19}$$

Further on, it will be of interest that $\delta_{\alpha\alpha}$ has the value

$$\delta_{\alpha\alpha} = 2 . \tag{20}$$

The boundary-part of eq. (16) reads under these conditions :

$$\oint^{B} \left[- D w_{,\alpha\alpha\nu} - P \right] \bar{w}\, ds +$$

$$+ \oint^{B} \Big[D \left(w_{,\alpha\beta} t_\alpha \nu_\beta + \nu w_{,\alpha\alpha} t_\beta \nu_\beta - \nu w_{,\alpha\beta} t_\beta \nu_\alpha \right) -$$

$$- \epsilon_{3\alpha\beta} \hat{M}_\alpha t_\beta \Big] \bar{w}_{,t} \, ds +$$

$$+ \oint^B \Big[D \left(w_{,\alpha\beta} \nu_\alpha \nu_\beta + \nu w_{,\alpha\alpha} \nu_\beta \nu_\beta - \nu w_{,\alpha\beta} \nu_\beta \nu_\alpha \right) -$$

$$- \epsilon_{3\alpha\beta} \hat{M}_\alpha \nu_\beta \Big] \bar{w}_{,\nu} \, ds = 0$$

or

$$\oint^B \left[-D w_{,\alpha\alpha\nu} - \hat{P} \right] \bar{w} \, ds +$$

$$+ \oint^B \Big[D \left(w_{,\alpha\beta} t_\alpha \nu_\beta + 0 - \nu w_{,\alpha\beta} t_\beta \nu_\alpha \right) - \epsilon_{3\alpha\beta} \hat{M}_\alpha t_\beta \Big] \bar{w}_{,t} \, ds +$$

$$+ \oint^B \Big[D \left(w_{,\alpha\beta} \nu_\alpha \nu_\beta + \nu w_{,\alpha\alpha} - \nu w_{,\alpha\beta} \nu_\alpha \nu_\beta \right) -$$

$$- \epsilon_{3\alpha\beta} \hat{M}_\alpha \nu_\beta \Big] \bar{w}_{,\nu} \, ds = 0 .$$

We make use of

$$\oint^B f \, \bar{w}_{,t} \, ds = \oint^B d \left(f \bar{w} \right) - \oint^B f_{,t} \, \bar{w} \, ds = 0 - \oint^B f_{,t} \, \bar{w} \, ds .$$

In this way, we get

$$\oint^B \left\{ - D w_{,\alpha\alpha\nu} - \hat{P} - \frac{\partial}{\partial s} \left[D \left(w_{,\alpha\beta} t_\alpha \nu_\beta - \nu w_{,\alpha\beta} t_\beta \nu_\alpha \right) - \epsilon_{3\alpha\beta} \hat{M}_\alpha t_\beta \right] \right\} \bar{w} \, ds +$$

$$+ \oint^B \Big\{ D \left[w_{,\alpha\beta} \nu_\alpha \nu_\beta + \nu \left(w_{,\alpha\alpha} - w_{,\alpha\beta} \nu_\alpha \nu_\beta \right) \right] -$$

$$- \epsilon_{3\alpha\beta} \hat{M}_\alpha \nu_\beta \Big\} \bar{w}_{,\nu} ds = 0 .$$

On the boundary, $\bar{w}$ and $\bar{w}_{,\nu}$ are independent from each other. It depends on the problem whether $\bar{w}$ and $\bar{w}_{,\nu}$ are arbitrary or have to be zero. At parts where w is not prescribed we get the natural boundary condition $(w_{,\alpha\beta} = w_{,\beta\alpha})$

$$- D w_{,\alpha\alpha\nu} - \hat{P} - \frac{\partial}{\partial s} \Big[D \left(w_{,\alpha\beta} t_\alpha \nu_\beta - \nu w_{,\alpha\beta} t_\alpha \nu_\beta \right) -$$

$$- \epsilon_{3\alpha\beta} \hat{M}_\alpha t_\beta \Big] = 0 .$$

$$D \left[w_{,\alpha\beta} \nu_\alpha \nu_\beta + \nu \left(w_{,\alpha\alpha} - w_{,\alpha\beta} \nu_\alpha \nu_\beta \right) \right] - \epsilon_{3\alpha\beta} \hat{M}_\alpha \nu_\beta = 0 .$$

These equations can be simplified by introducing $\hat{M}_\nu = \hat{M}_\alpha \nu_\alpha$ and $\hat{M}_t = \hat{M}_\alpha t_\alpha$. We see that this yields with eqs (19) :

$$\hat{M}_\nu = \hat{M}_\alpha \nu_\alpha = \hat{M}_\alpha \epsilon_{3\alpha\beta} t_\beta \quad ; \quad \hat{M}_t = - \hat{M}_\alpha \epsilon_{3\alpha\beta} \nu_\beta .$$

The eqs (19) help us to simplify $w_{,\alpha\alpha} - w_{,\alpha\beta} \nu_\alpha \nu_\beta$:

$$w_{,\alpha\alpha} - w_{,\alpha\beta} \nu_\alpha \nu_\beta = w_{,\alpha\alpha} - w_{,\alpha\beta} \epsilon_{3\alpha\gamma} t_\gamma \epsilon_{3\beta\delta} t_\delta =$$

$$= w_{,\alpha\alpha} - w_{,\alpha\beta} t_\gamma t_\delta \left(\delta_{\alpha\beta} \delta_{\gamma\delta} - \delta_{\alpha\delta} \delta_{\beta\gamma} \right) =$$

$$= w_{,\alpha\alpha} - w_{,\alpha\alpha} t_\gamma t_\gamma + w_{,\alpha\beta} t_\alpha t_\beta = w_{,\alpha\beta} t_\alpha t_\beta .$$

The boundary equations now read :

$$\hat{P} - \frac{\partial}{\partial s}\hat{M}_{\nu} + D\left[\frac{\partial \Delta w}{\partial \xi} + (1-\nu)\frac{\partial}{\partial s}\left(w_{,\alpha\beta}\, t_{\alpha}\nu_{\beta}\right)\right] = 0 ,$$

$$\hat{M}_t + D\, w_{,\alpha\beta}\left(\nu_{\alpha}\nu_{\beta} + \nu\, t_{\alpha} t_{\beta}\right) = 0 .$$

The most interesting aspect of this result is that we cannot distinguish between $\hat{P}$ and the derivative of the torsion moment $\hat{M}_{\nu}$.

12.2. Theorems of Cotteril Castigliano

In this chapter we will first apply the two theorems of COTTERIL-CASTIGLIANO to a simple problem of beam theory. Afterwards, we will try to apply the upper- and lower-bound theorems to these methods .

Problem 12.2-1 : Find out the displacement of the beam of Fig. 12.2-1 at that point where $\hat{P}$ is acting as a function of $\hat{P}$ by means of the two theorems of COTTERIL-CASTIGLIANO. Assume pure bending.

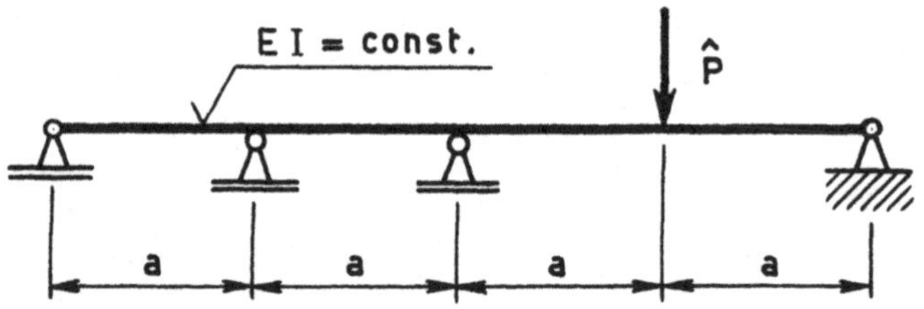

Fig. 12.2-1

I) 2nd theorem of CASTIGLIANO

The formula belonging to this theorem is

$$\hat{x}_i = \left.\frac{\partial V(Q)}{\partial F_i}\right|_{\hat{F}} , \quad \text{where} \quad V(Q) = \frac{1}{2}\int \frac{M_b^2}{EI}\,d\xi .$$

In order to make the integration easier we introduce a "normal part" of $M_b(\xi)$ (ξ = beam-coordinate). In this part of the beam, M_b depends linearly on ξ (cf. Fig. 12.2-2)

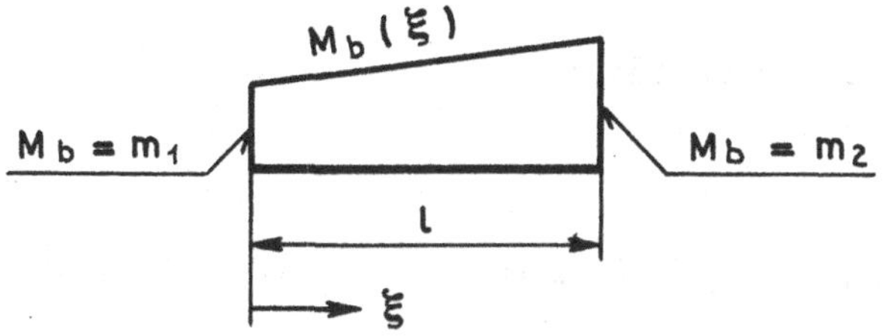

Fig 12.2-2.

To the normal part sketched in Fig 12 2-2, a $\Delta V(Q)$ belongs which is given by

$$\Delta Q(V) = \frac{\ell}{GEI}\left[m_1^2 + m_1 m_2 + m_2^2\right].$$

In our case, the distribution of the bending moment can easily be calculated when the system is made statically determinate by making it free at two supports and by introducing the corresponding reaction forces at these points. These forces are handled as external loads. Such a statically determinate system is given at Fig. 12.2-3 where $M_b(\xi)$ is printed, too.

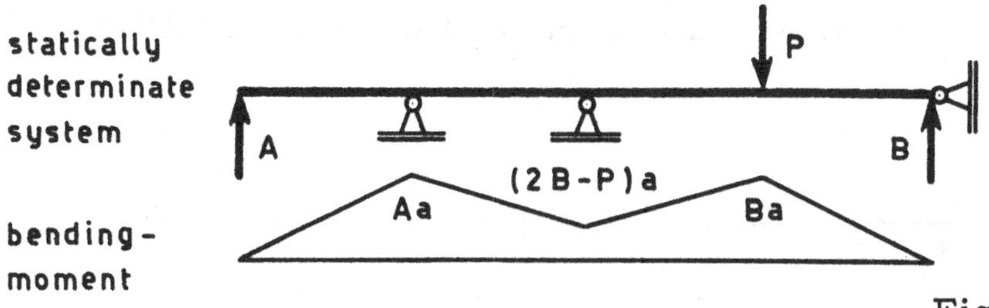

Fig 12.2-3

Integrating $V(Q)$ we reach

$$V(Q) = \frac{a^3}{6EI}\left\{2A^2 + 12B^2 + 2P^2 + 2AB - AP - 9BP\right\}.$$

The basic equation yields :

$$\hat{x}_A = \left.\frac{\partial V}{\partial A}\right|_{\hat{F}} = 0 \quad : \qquad 4\hat{A} + 2\hat{B} = \hat{P},$$

$$\hat{x}_B = \left.\frac{\partial V}{\partial B}\right|_{\hat{F}} = 0 \quad : \qquad 2\hat{A} + 24\hat{B} = 9\hat{P}.$$

From these relations we get : $\hat{A} = \frac{3}{46}\hat{P}$; $\hat{B} = \frac{17}{46}\hat{P}$.

$$\hat{x}_P = \left.\frac{\partial V}{\partial P}\right|_{\hat{F}} = \frac{a^3}{6EI}\left\{4\hat{P} - \hat{A} - 9\hat{B}\right\} = \frac{7}{69}\frac{\hat{P}a^3}{EI}.$$

is the result.

II) First theorem of CASTIGLIANO

The basic formula of the first theorem of CASTIGLIANO is

$$\hat{F}_j = \left.\frac{\partial U(q)}{\partial x_j}\right|_{\hat{q}} \quad \text{where} \quad U(q) = \frac{1}{2}\int EI\,(w'')^2\,d\xi .$$

To reach a true solution we must be able to express the true values of q by the values of a small number of external displacements x_j which must not be known from the beginning. In our case we know from the statics (!) that $w''(\xi)$ is a linear function by the piece In such a case we can introduce a "normal part" of the integration as we did at the second theorem. The normal part is printed in Fig 12.2-4. The potential energy belonging to it is

$$\Delta U(q) = \frac{2EI}{\ell^3}\left\{3\,(w_1 - w_2)^2 + 3\ell\,(w_1 - w_2)(\psi_1 + \psi_2) + \ell^2(\psi_1^2 + \psi_1\psi_2 + \psi_2^2)\right\}.$$

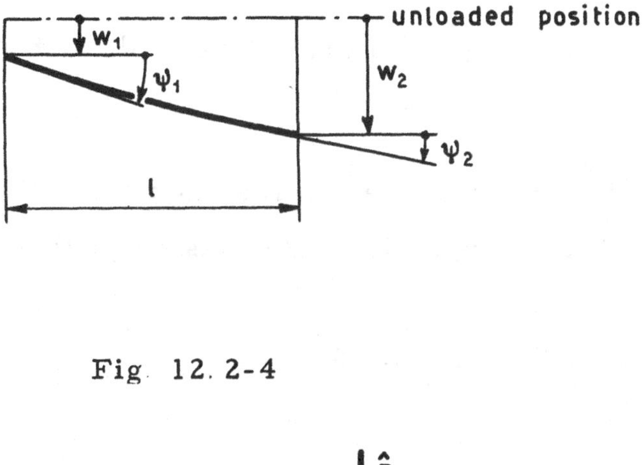

Fig 12.2-4

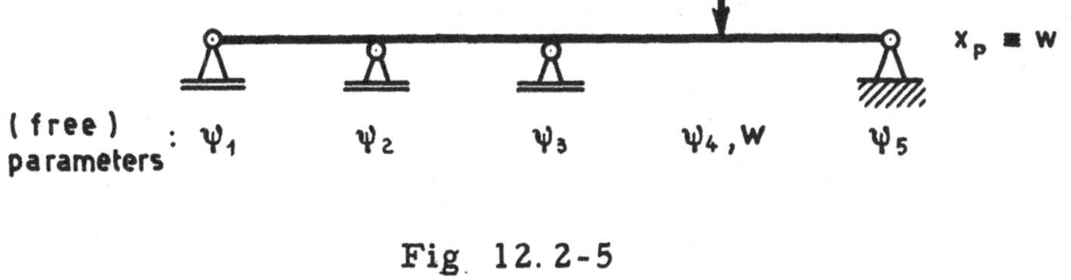

Fig 12.2-5

The parameters which are necessary for the description of the internal displacements w'', are ψ_1 to ψ_5 and w (cf. Fig. 12.2-5) Integrating $U(q)$ by means of four normal parts we get

$$U(q) = \frac{2EI}{a^3} \cdot \Big\{ a^2(\psi_1^2 + \psi_1\psi_2 + \psi_2^2) + a^2(\psi_2^2 + \psi_2\psi_3 + \psi_3^2) +$$

$$+ 3w^2 - 3aw(\psi_3 + \psi_4) + a^2(\psi_3^2 + \psi_3\psi_4 + \psi_4^2) +$$

$$+ 3w^2 + 3aw(\psi_4 + \psi_5) + a^2(\psi_4^2 + \psi_4\psi_5 + \psi_5^2)\Big\} =$$

$$= \frac{2EI}{a^3} \Big\{ 6w^2 + 3aw(\psi_5 - \psi_3) + a^2(\psi_1^2 + \psi_1\psi_2 + 2\psi_2^2 + \psi_2\psi_3 +$$

$$+ 2\psi_3^2 + \psi_3\psi_4 + 2\psi_4^4 + \psi_4\psi_5 + \psi_5^2)\Big\}.$$

With $\hat{M}_i$ being the true external moment at that point where the inclination angle is ψ_i, the basic equation yields

$$\left.\frac{\partial U}{\partial \psi_1}\right|_{\hat{\psi},\hat{w}} = \hat{M}_1 = 0 \Longrightarrow 2\hat{\psi}_1 + \hat{\psi}_2 \quad \cdot \quad \cdot \quad \cdot \quad \cdot = 0,$$

$$\left.\frac{\partial U}{\partial \psi_2}\right|_{\hat{\psi},\hat{w}} = \hat{M}_2 = 0 \Longrightarrow \hat{\psi}_1 + 4\hat{\psi}_2 + \hat{\psi}_3 \quad \cdot \quad \cdot \quad \cdot = 0,$$

$$\left.\frac{\partial U}{\partial \psi_3}\right|_{\hat{\psi},\hat{w}} = \hat{M}_3 = 0 \Longrightarrow \cdot \quad \hat{\psi}_2 + 4\hat{\psi}_3 + \hat{\psi}_4 \quad \cdot \quad -3\frac{\hat{w}}{a} = 0,$$

$$\left.\frac{\partial U}{\partial \psi_4}\right|_{\hat{\psi},\hat{w}} = \hat{M}_4 = 0 \implies \quad \cdot \quad \cdot \quad \hat{\psi}_3 + 4\hat{\psi}_4 + \hat{\psi}_5 \quad \cdot \quad = 0,$$

$$\left.\frac{\partial U}{\partial \psi_5}\right|_{\hat{\psi},\hat{w}} = \hat{M}_5 = 0 \implies \quad \cdot \quad \cdot \quad \cdot \quad \hat{\psi}_4 + \hat{\psi}_5 + 3\frac{w}{a} = 0,$$

$$\left.\frac{\partial U}{\partial w}\right|_{\hat{\psi},\hat{w}} = \hat{P} \neq 0 \implies \quad \cdot \quad \cdot \quad -\hat{\psi}_3 \quad \cdot \quad + \hat{\psi}_5 + \frac{\hat{w}}{a} = \frac{\hat{P}a^2}{6EI}.$$

The solution of this set of equations is

$$\hat{w} = \frac{7}{69}\,\frac{\hat{P}a^3}{EI}$$

III) An upper bound for $\hat{P}$

We start our considerations from $U(q) \geq U(\hat{q})$ if $\Lambda_{ex}(\hat{F}<\hat{x}>) = \Lambda_{ex}(\hat{F}<x>)$. Applying the first theorem of CASTIGLIANO, we prescribe $w = \hat{w}$. On the other side, $\hat{P}$ is the only effective external load Therefore $\Lambda_{ex}(\hat{F}<\hat{x}>)$ is identical with $\hat{P}\hat{w} = \hat{P}w = \Lambda_{ex}(\hat{F}<x>)$. Hence, $U(q) \geq U(\hat{q})$ is valid for every curve $w(\xi)$ satisfying $w = \hat{w}$ and the geometrical boundary conditions.

In the case of linear theory, $U(q)$ can be expressed by the external loads to be equal to

$$U(q) = \frac{1}{2}\Lambda_{ex}(F < x >) = \frac{1}{2} P w + \frac{1}{2}\sum_i F_i^r x_i^r =$$

$$= \frac{1}{2} P \hat{w} + \frac{1}{2}\sum_i F_i^r x_i^r .$$

Where F_i^r stands for an external force (except $\hat{P}$) which belongs to a true solution where $U(q)$ is the true potential. These forces F_i would be calculated by the first theorem of CASTIGLIANO if we use this not-true potential U. If $\sum_i F_i^r x_i^r = 0$, we get

$$\frac{1}{2} P \hat{w} \geq \frac{1}{2} \hat{P} \hat{w} \quad \text{or} \quad \underline{\underline{P \geq \hat{P}}} .$$

On the other hand, $F_i x_i = 0$ is valid if we use systems where $F_i = 0$ or $x_i = 0$. In the true solution we always had $\hat{x}_i = \hat{\psi}_i \neq 0$, but $\hat{F}_i = \hat{M}_i = 0$. If we take $\psi_i = 0$ instead of $M_i = 0$ we get an upper bound for $\hat{P}$.

So it is possible to get upper bounds for $\hat{P}$ by applying the first theorem of CASTIGLIANO when using parameters x_i as to be zero, which in reality are free. This result is physically expectable : we make the system stiffer. An example for this upper bound theory can be found if we say

$$\psi_2 = 0 :$$

$U\,(q)$ now gets :

$$U(q) = \frac{2EI}{a^3}\left\{6w^2 + 3aw\,(\psi_5 - \psi_3) + \right.$$

$$\left. + a^2(\psi_1^2 + 2\psi_3^2 + \psi_3\psi_4^2 + 2\psi_4^2 + \psi_4\psi_5 + \psi_5^2)\right\}.$$

This leads to :

$$\frac{\partial U}{\partial \psi_1} = M_1 = 0 = \frac{2EI}{a^3}\cdot 2\psi_1 a^2 \implies \psi_1 = 0$$

$$\frac{\partial U}{\partial \psi_3} = M_3 = 0 = \frac{2EI}{a^3}\left\{-3aw + a^2(4\psi_3 + \psi_4)\right\} = 0$$

$$\frac{\partial U}{\partial \psi_4} = M_4 = 0 = \frac{2EI}{a^3}\left\{a^2(\psi_3 + 4\psi_4 + \psi_5)\right\} = 0$$

$$\frac{\partial U}{\partial \psi_5} = M_5 = 0 = \frac{2EI}{a^3}\left\{3aw + a^2(\psi_4 + 2\psi_5)\right\} = 0$$

$$\frac{\partial U}{\partial w} = P = \frac{2EI}{a^3}\left\{12w + 3a(\psi_5 - \psi_3)\right\}$$

The result of this system of equations is :

$$\hat{P} \leq P = \frac{132}{13}\,\frac{EI}{a^3} \quad \text{instead of} \quad \hat{P} = \frac{69}{7}\,\frac{EI}{a^3}\,.$$

We recognize :

$$P - \hat{P} = \frac{132 \cdot 7 - 69 \cdot 13}{7 \cdot 13} \frac{EI}{a^3} = \frac{27}{7 \cdot 13} \frac{EI}{a^3} > 0 .$$

IV) An upper bound for $\hat{w}$

The discussion will be similar to the considerations in connection with the upper bound for $\hat{P}$:

$$V(Q) \geq V(\hat{Q}) \quad \text{holds if} \quad \Lambda_{ex}(F < \hat{x} >) = \Lambda_{ex}(\hat{F} < \hat{x} >) .$$

Applying the second theorem of CASTIGLIANO we prescribe the external effective forces $\hat{F}$ so that $V(Q) \geq V(\hat{Q})$ is given. $V(Q)$ is equal to $\frac{1}{2} P w$ if $\sum_i F_i^r x_i^r = 0$. The products $F_i^r x_i^r$ vanish if $x_i^r = 0$, this yields the true result. But, possibly, F_i^r vanish which in reality are free. Then, we state

$$P w = \hat{P} w \geq \hat{P} \hat{w} \quad \text{or} \quad w \geq \hat{w} .$$

This yields an upper bound for $\hat{w}$, or if we think of w to be given, a lower bound for $\hat{P}$.

This means : If we take a reaction force to be zero which in reality had to be used as an external unknown load on the system being made statically determinate, whereas really $x_i = 0$ had to be valid at this point, and calculate $V(Q)$ only by means of the remaining forces we get an upper bound for $\hat{w}$.

At our example $A = 0$, $B \neq 0$ yields :

$$V(Q) = \frac{a^3}{6EI}\left(12B^2 + 2P^2 - 9BP\right)$$

$$\frac{\partial V}{\partial B} = x_B = 0 \implies 24B - 9P = 0 \ , \ B = \frac{3}{8}P$$

$$\frac{\partial V}{\partial P} = w = \frac{a^3}{6EI}(4P - 9B) = \frac{Pa^3}{6EI}\cdot\left(4 - \frac{27}{8}\right) = \frac{5}{48}\frac{Pa^3}{EI}$$

instead of $\hat{w} = \frac{7}{69}\frac{\hat{P}a^3}{EI}$:

$$w - \hat{w} = \left(\frac{5}{48} - \frac{7}{69}\right)\frac{\hat{P}a^3}{EI} = \frac{345 - 336}{48\cdot 69}\frac{\hat{P}a^3}{EI} > 0 .$$

The simplification would have been too drastic if we would have said $A = B = 0$:

$$V(Q) = \frac{P^2a^3}{3EI} \ , \quad w = \frac{\partial V}{\partial P} = \frac{2}{3}\frac{Pa^3}{EI} = \frac{46}{69}\frac{Pa^3}{EI} ,$$

but it would have lead to an upper bound for $\hat{w}$, too .

12.3. Torsion of prismatic beams

In this section, upper- and lower-bound methods will be applied to the torsion of prismatic beams with cross-sections which have at least two axes of symmetry. Thus, we consider beams with cross-section as they are sketched in Fig 12. 3-1

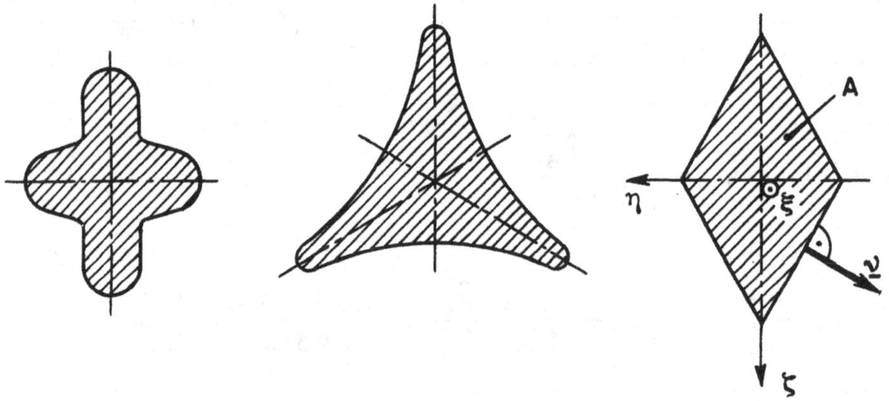

Fig. 12. 3-1

The at least double symmetry is necessary because we need a fixed axis in space which is a fixed axis in the beam simultaneously. We use the coordinates shown in Fig. 12. 3-1.

I) True solution

We start our considerations from the following assumption :

If a prismatic beam is twisted, the cross-section moves nearly like a rigid body. Its projection on a η,ζ-plane rotates like a rigid body. Additional displacements are possible in ξ-direction and will be called u.

The torsion-angle per unit length (ξ) of the beam will be denoted by ϑ. For a problem of this kind, where the torsion-moment $\hat{M}_t$ is given, ϑ and the displacements $u(\eta,\zeta)$ are free variables. We assume : $M_t \neq M_t(\xi)$, $u \neq u(\xi)$. Under these conditions the displacements u (in ξ - direction) v (in η - direction) and w (in ζ - direction) are :

$$u = u(\zeta,\eta)\ ; \quad v = -\vartheta\,\xi\,\zeta\ ; \quad w = \vartheta\,\xi\cdot\eta\ . \tag{1}$$

If $M_t = \text{const.}$, it is obvious to state $\vartheta = \text{const.}$ The (linear) deformations ε_{ij} can be calculated from relations (1) :

$$[\varepsilon_{ij}] = \begin{bmatrix} 0 & \frac{1}{2}(u_{,\eta} - \vartheta\zeta) & \frac{1}{2}(u_{,\zeta} + \vartheta\eta) \\ \frac{1}{2}(u_{,\eta} - \vartheta\zeta) & 0 & 0 \\ \frac{1}{2}(u_{,\zeta} + \vartheta\eta) & 0 & 0 \end{bmatrix}.$$

Now we are enabled to establish the theorem of the minimum of the potential energy :

$$U(q) - \hat{M}_t \vartheta \ell \Longrightarrow \min .$$

This becomes in our case by

$$U(q) = \ell \iint^{A} G(\varepsilon^2_{\xi\eta} + \varepsilon^2_{\xi\zeta})\, dA :$$

$$F = \frac{1}{2} \iint^{A} G\left[(u_{,\eta} - \vartheta\zeta)^2 + (u_{,\zeta} + \vartheta\eta)^2\right] dA -$$

$$- \hat{M}_t \vartheta \Longrightarrow \min . \tag{2}$$

This principle will lead us to a best approximation of the true result. First we are interested in the EULER-LAGRANGE-equations and the boundary conditions satisfied by the optimal solution to relation (2). Introducing $u = \tilde{u} + \varepsilon\bar{u}$, $\vartheta = \tilde{\vartheta} + \varepsilon\bar{\vartheta}$, we get from $\partial F / \partial\varepsilon = 0$

$$\iint^{A} G\left[(\tilde{u}_{,\eta} - \tilde{\vartheta}\zeta)(\bar{u}_{,\eta} - \bar{\vartheta}\zeta) + \right.$$

$$\left. + (\tilde{u}_{,\zeta} + \tilde{\vartheta}\eta)(\bar{u}_{,\zeta} + \bar{\vartheta}\eta)\right] dA - \hat{M}_t \bar{\vartheta} = 0$$

or

$$G \oint^{B} \left[(\tilde{u}_{,\eta} - \tilde{\vartheta}\zeta)\nu_\eta + (\tilde{u}_{,\zeta} + \tilde{\vartheta}\eta)\nu_\zeta\right] \bar{u}\, dB -$$

$$- G \iint^{A} \left[\tilde{u}_{,\eta\eta} + \tilde{u}_{,\zeta\zeta}\right] \bar{u}\, dA +$$

$$+ \bar{\vartheta} \cdot \left\{ G \iint^{A} \left[\zeta(\tilde{\vartheta}\zeta - \tilde{u}_{,\eta}) + \eta(\tilde{\vartheta}\eta + \tilde{u}_{,\zeta})\right] dA - \hat{M}_t \right\} = 0 .$$

This leads to the following relations :

$$(\tilde{u}_{,\eta} - \tilde{\vartheta}\zeta)\,\nu_\eta + (\tilde{u}_{,\zeta} + \tilde{\vartheta}\eta)\,\nu_\zeta = 0 \quad \text{at the boundary,} \tag{3}$$

$$\tilde{u}_{,\eta\eta} + \tilde{u}_{,\zeta\zeta} = 0 \quad \text{in the interiour ,} \tag{4}$$

and :

$$\tilde{M}_t = G \iint^{A} \left[\zeta\,(\tilde{\vartheta}\zeta - \tilde{u}_{,\eta}) + \eta\,(\tilde{\vartheta}\eta + \tilde{u}_{,\zeta}) \right] dA \,.$$

The last relation is not so interesting for us.

We want to show that the solution $\tilde{u}$, $\tilde{\vartheta}$ of the problem (2) is a correct solution $\hat{u}$, $\hat{\vartheta}$ for torsion problems and not only an approximate solution as we should expect, because we started from arbitrary assumptions.

1. The compatibility conditions are satisfied because we used u, v, w instead of ε_{ij}.

2. The stresses which belong to our assumptions are

$$[\sigma_{ij}] = \begin{bmatrix} 0 & G(u_{,\eta} - \vartheta\zeta) & G(u_{,\zeta} + \vartheta\eta) \\ G(u_{,\eta} - \vartheta\zeta) & 0 & 0 \\ G(u_{,\zeta} + \vartheta\eta) & 0 & 0 \end{bmatrix} .$$

This satisfies the constitutive law.

3. The last relations which have to be satisfied in the interior

are the equations of equilibrium : $\sigma_{ij,i} = 0$, they yield:

$$\sigma_{\xi\xi,\xi} + \sigma_{\eta\xi,\eta} + \sigma_{\zeta\xi,\zeta} = 0 + Gu_{,\eta\eta} + Gu_{,\zeta\zeta} \stackrel{!}{=} 0,$$

this is just eq. (4),

$$\sigma_{\xi\eta,\xi} + \sigma_{\eta\eta,\eta} + \sigma_{\zeta\eta,\zeta} = 0 + 0 + 0 \stackrel{!}{=} 0 \quad \text{(satisfied)}$$

$$\sigma_{\xi\zeta,\xi} + \sigma_{\eta\zeta,\eta} + \sigma_{\zeta\zeta,\zeta} = 0 + 0 + 0 \stackrel{!}{=} 0 \quad \text{(satisfied)}.$$

Thus, in the interior, each relation is satisfied by the result $\tilde{u}_i$, $\tilde{\vartheta}$.

4. The last question is whether also the boundary-conditions $\sigma_{\nu i} = 0$ or $\sigma_{ki}\nu_k = 0$ are valid. We know that $\nu_\xi = 0$. Thus, we get :

$$\sigma_{\xi\xi}\nu_\xi + \sigma_{\eta\xi}\nu_\eta + \sigma_{\zeta\xi}\nu_\zeta =$$
$$= 0 + G\left[(u_{,\eta} - \vartheta\zeta)\nu_\eta + (u_{,\zeta} + \vartheta\eta)\nu_\zeta\right] \stackrel{!}{=} 0 ,$$

which is just eq. (3), and :

$$\sigma_{\xi\eta}\nu_\xi + \sigma_{\eta\eta}\nu_\eta + \sigma_{\xi\eta}\nu_\zeta = 0 + 0 + 0 \stackrel{!}{=} 0 \quad \text{(satisfied)},$$

$$\sigma_{\xi\zeta}\nu_\xi + \sigma_{\eta\zeta}\nu_\eta + \sigma_{\zeta\zeta}\nu_\zeta = 0 + 0 + 0 \stackrel{!}{=} 0 \quad \text{(satisfied)}.$$

We see that the result of our considerations is a correct solution of the torsion problem. In fact, the displacements $u = u(\eta, \zeta)$ must be free, hence, we have to assume that the considered cross-section is far enough from the ends of the beam.

II) Upper bound for $\hat{I}_t$.

Assuming $\vartheta = \hat{\vartheta}$, we can state

$$U(q,\hat{\vartheta}) - \hat{M}_t \hat{\vartheta}_\ell \geq U(\hat{q},\hat{\vartheta}) - \hat{M}_t \hat{\vartheta}_\ell$$

or

$$\frac{1}{2}\ell \iint^A G\left[(u_{,\eta} - \hat{\vartheta}\zeta)^2 + (u_{,\zeta} + \hat{\vartheta}\eta)^2\right] dA \geq U(\hat{q}) - \frac{1}{2}\hat{M}_t \hat{\vartheta}\ell \,.$$

Dividing by $\frac{1}{2} G\ell \hat{\vartheta}^2$ we get with $z(\eta,\zeta) = u(\eta,\zeta)/\hat{\vartheta}$:

$$\iint^A \left[(z_{,\eta} - \zeta)^2 + (z_{,\zeta} + \eta)^2\right] dA \geq \frac{\hat{M}_t}{G\hat{\vartheta}} = \hat{I}_t \,.$$

Thus, we have derived an upper bound theorem for $\hat{I}_t$. z is an arbitrary function of η and ζ. The theorem is :

$$\hat{I}_t = \min\left(\iint^A \left[(z_{,\eta} - \zeta)^2 + (z_{,\zeta} + \eta)^2\right] dA\right). \qquad (5)$$

The boundary conditions for the true solution are :

$$(z_{,\eta} - \zeta)\nu_\eta + (z_{,\zeta} + \eta)\nu_\zeta = 0 \qquad \text{(cf. eq. (3))}. \qquad (6)$$

III) Lower bound for $\hat{I}_t$

To find out lower bounds we have to start our considerations from

$$U(\hat{q}) \geq \frac{1}{4} \frac{\Lambda_{ex}(F < \hat{x} >)}{V(Q)},$$

where $U(\hat{q}) = \frac{1}{2} \Lambda_{ex}(\hat{F} < \hat{x} >) = \frac{1}{2} \hat{M}_t \hat{\vartheta} \ell = \frac{1}{2} G \hat{I}_t \hat{\vartheta}^2 \ell .$

We know that the true solution contains $\tau_\eta = \sigma_{\xi\eta}$, $\tau_\zeta = \sigma_{\xi\zeta}$ and no other function. Therefore $V(Q)$ is given by

$$V(Q) = \frac{\ell}{2G} \iint^{A} (\tau_\eta^2 + \tau_\zeta^2)\, dA .$$

Λ_{ex} is equal to the torsion moment M_t multiplied by $\hat{\vartheta}\ell$ if and only if there are no shear stresses acting at the surface of the beam which could act against the displacements, and if no volume forces belong to the guessed stress-state. Therefore an admissible stress-field which can advantageously be used has to satisfy

$$\sigma_{ij,i} = 0 .$$

This yields only one relation :

(7) $$\tau_{\eta,\eta} + \tau_{\zeta,\zeta} = 0 .$$

The boundary condition is

$$\tau_\eta \nu_\eta + \tau_\zeta \nu_\zeta = 0 . \tag{8}$$

A limit-case of eq. (7) is a jump of τ_1, parallel to a jump-line (cf. Fig 12. 3-2).

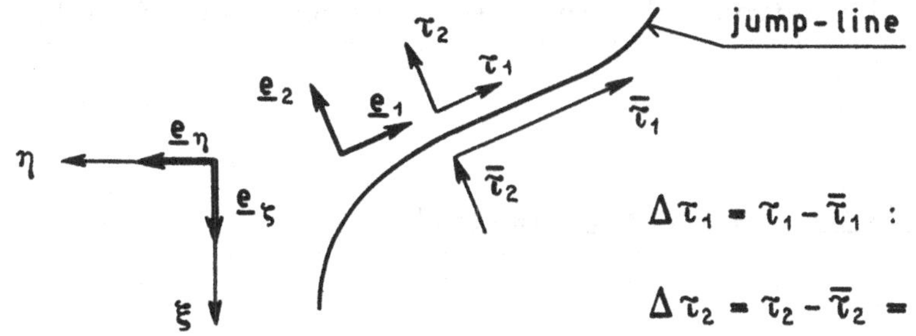

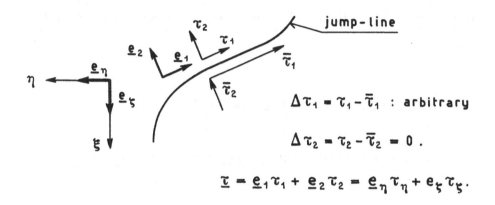

$$\Delta\tau_1 = \tau_1 - \bar{\tau}_1 \ : \ \text{arbitrary}$$

$$\Delta\tau_2 = \tau_2 - \bar{\tau}_2 = 0 .$$

$$\underline{\tau} = \underline{e}_1 \tau_1 + \underline{e}_2 \tau_2 = \underline{e}_\eta \tau_\eta + e_\zeta \tau_\zeta .$$

Fig 12. 3-2

If we have estimated such a field of stresses we can calculate a lower bound for $\hat{I}_t$ by

$$\frac{1}{2} G \hat{\vartheta}^2 \hat{I}_t \ell \geq \frac{1}{4} \frac{\ell^2 \hat{\vartheta}^2 \left[\iint^A (\tau_\eta \zeta - \tau_\zeta \eta) \, dA \right]^2}{\frac{\ell}{2G} \iint^A (\tau_\eta^2 + \tau_\zeta^2) \, dA}$$

or

(9) $$\hat{I}_t \geq \frac{\left[\iint\limits^{A} (\tau_\eta \zeta - \tau_\zeta \eta)\, dA\right]^2}{\iint\limits^{A} (\tau_\eta^2 + \tau_\zeta^2)\, dA}$$

But we can imagine that it is not easy to calculate good lower bounds because the estimated stress fields have to satisfy a lot of conditions.

IV) Lower bounds for a square cross-section

a) We estimate the stress field of Fig. 12. 3-3 ($\underline{\tau}$: cf. Fig. 12. 3-2).

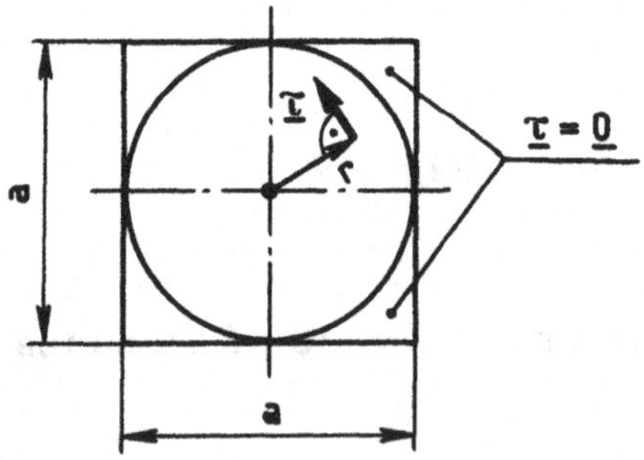

Fig. 12. 3-3

We state : $$|\underline{\tau}| = \tau(r) = A + Br \quad \text{if} \quad r \leq \frac{a}{2},$$
$$|\underline{\tau}| = 0 \quad \text{if} \quad r > \frac{a}{2}.$$

We will not check here that eqs. (7) and (8) are satisfied. The cyrcle $r = \frac{a}{2}$ is an admissible jump-line (cf. Fig. 12. 3-2).

When integrating we get

$$\hat{I}_t \geq \frac{\left[\int_0^{\frac{a}{2}} 2\pi r(A+Br)\, r\, dr\right]^2}{\int_0^{\frac{a}{2}} 2\pi r\,(A+Br)^2 dr} = \pi\,\frac{a^6}{a^2}\cdot\frac{\left(A\,\frac{1}{12}+Ba\cdot\frac{1}{32}\right)^2}{\left(A^2\cdot\frac{1}{4}+ABa\cdot\frac{1}{6}+B^2a^2\cdot\frac{1}{32}\right)} =$$

$$= \pi\,a^4\cdot\frac{1}{96}\cdot\frac{(8+3\alpha)^2}{24+16\,\alpha+3\,\alpha^2}\,,$$

where $\alpha = \frac{Ba}{A}$.

The maximization of $g = \frac{(8+3\alpha)^2}{24+16\,\alpha+3\,\alpha^2}$ yields

$$6\,(8+3\,\alpha)(24+16\,\alpha+3\,\alpha^2)-(8+3\alpha)^2(16+6\alpha) = 0$$

or

$$(144+\cancel{96\alpha}+\cancel{18\alpha^2}-128-\cancel{96\alpha}-\cancel{18\alpha^2})\,(8+3\,\alpha) = 16\,(8+3\alpha) = 0\,.$$

We only get one extremum : $8+3\,\alpha = 0$ with $g = 0$.

This is a minimum : The maximum belongs to $\alpha \longrightarrow \infty$:

$g_{opt} = 3$.

Thus, we find out

$$\hat{I}_t \geq \pi\,\frac{a^4}{32} \approx 0.1\,a^4\,,$$

which is the well-known result for circular cross-sections.

b)

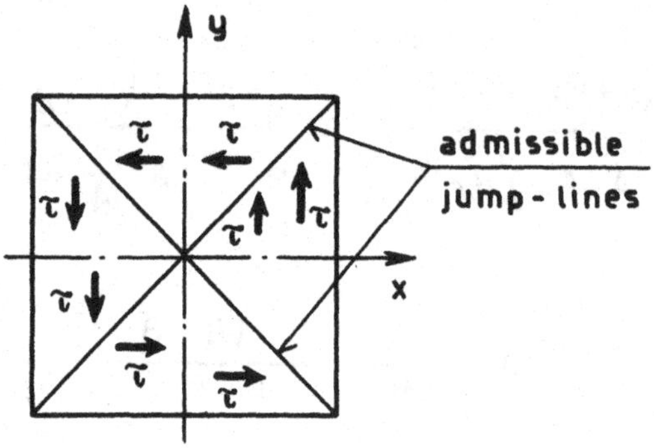

Fig. 12.3-4

In the area $x > 0$, $y < x$, $y > -x$, τ has the form :

$$\tau_y = A\left(1 + \alpha \frac{x}{a}\right); \quad \tau_x = 0 .$$

This approximation yields

$$\hat{I}_t \geq \frac{a^4}{24} \, \frac{64 + 48\alpha + 9\alpha^2}{24 + 16\alpha + 3\alpha^2} ,$$

$$\alpha_{opt} \longrightarrow \infty .$$

$\alpha = 0$ leads to

$$\hat{I}_t \geq \frac{a^4}{9} = 0.1111\, a^4 ,$$

$\alpha = \infty$ yields the - up to now - best result :

$$\hat{I}_t \geq \frac{a^4}{8} = 0.125\, a^4 .$$

c) Introduction of a stress-function

If we introduce a stress function by

$$\tau_\eta = +\Phi_{,\xi} \,, \quad \tau_\xi = -\Phi_{,\eta} \,,$$

eq. (7) is always satisfied (Φ has to be twice differentiable), The boundary condition (8) then yields :

$$\Phi_{,\eta}\, \nu_\xi - \Phi_{,\xi}\, \nu_\eta = 0 \qquad \text{or}$$

$$\Phi_{,\eta}\, t_\eta + \Phi_{,\xi}\, t_\xi = 0 \qquad \text{or}$$

$$\Phi_{,t} = 0$$

($\underline{t}$: tangential unit vector at the boundary; $\Phi_{,t}$: derivative in direction of $\underline{t}$). This is satisfied if $\Phi = 0$ at the boundary, Then, our lower-bound theorem yields :

$$\hat{I}_t \geq \frac{\left[\iint^A (\Phi_{,\xi}\, \xi + \Phi_{,\eta}\, \eta)\, dA\right]^2}{\iint^A (\Phi_{,\xi}^2 + \Phi_{,\eta}^2)\, dA} \,,$$

where Φ is any arbitrary function which is zero at the boundary

Let us use

$$\Phi = \left[1-\left(\frac{2\eta}{a}\right)^2\right]\cdot\left[1-\left(\frac{2\zeta}{a}\right)^2\right]:$$

$$\Phi_{,\eta} = -\frac{8\eta}{a^2}\left[1-\left(\frac{2\zeta}{a}\right)^2\right]; \quad \Phi_{,\zeta} = -\frac{8\zeta}{a^2}\left[1-\left(\frac{2\eta}{a}\right)\right].$$

This leads to

$$\iint^A (\Phi_{,\zeta}\,\zeta + \Phi_{,\eta}\,\eta)\,dA =$$

$$= -8\int_0^{\frac{a}{2}}\int_0^{\frac{a}{2}}\left\{\left(\frac{2\eta}{a}\right)^2\left[1-\left(\frac{2\zeta}{a}\right)^2\right]+\left(\frac{2\zeta}{a}\right)^2\left[1-\left(\frac{2\eta}{a}\right)^2\right]\right\}d\eta\,d\zeta =$$

$$= -\frac{8}{9}\,a^2$$

and

$$\iint^A (\Phi_{,\eta}^2 + \Phi_{,\zeta}^2)\,dA = \iint^A 2\,\Phi_{,\eta}^2\,dA =$$

$$= 2\iint^A \frac{64\eta^2}{a^4}\left[1-2\left(\frac{2\zeta}{a}\right)^2+\left(\frac{2\zeta}{a}\right)^4\right]d\eta\,d\zeta =$$

$$= 8\int_0^{\frac{a}{2}}\int_0^{\frac{a}{2}} \frac{16}{a^2}\left(\frac{2\eta}{a}\right)^2\left[1-2\left(\frac{2\zeta}{a}\right)^2+\left(\frac{2\zeta}{a}\right)^4\right]d\eta\,d\zeta =$$

$$= 8\cdot\frac{4}{a^2}\cdot\frac{1}{3}\cdot\left[1-\frac{2}{3}+\frac{1}{5}\right]a^2 = \frac{32}{3}\cdot\frac{8}{15} = \frac{256}{45}.$$

Hence the result is

$$\hat{I}_t \geq \frac{45}{256} \cdot \left(\frac{8}{9} a^2\right)^2 = \frac{45 \cdot 64}{256 \cdot 81} a^4 = \frac{5}{36} a^4 = \underline{\underline{0.13889 a^4}}.$$

V) Upper bounds for a square cross-section

We start from eq. (5) where $z(\eta,\zeta)$ is an arbitrary function. There is no admissibility condition. If we use $z = 0$, the boundary condition (6) is not satisfied The result is

$$\hat{I}_t \leq (\zeta^2 + \eta^2)\, dA = I_p = \frac{1}{6} a^4 = 0.1667\, a^4 .$$

Thus, we know that $\hat{I}_t$ lies between $0.1389 a^4$ and $0.1667 a^4$. Let us try to find out a lower "upper bound". At first, we recognize that the axes I to IV of Fig. 12. 3-5 cannot move because of symmetry. We can state more : these axes must be axes of antisymmetry for $u(\eta,\zeta)$. This shows that the potential energy stored in one eighth of A (for example area ①) is 1/8 of the total potential energy

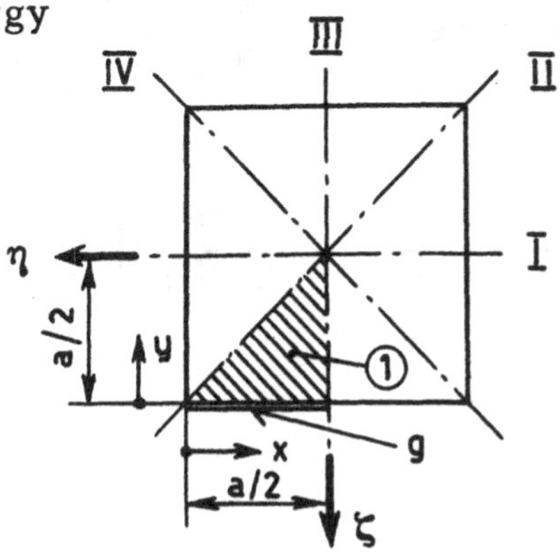

Fig. 12. 3-5

Hence we write

$$\hat{I}_t \leq 8 \overset{①}{\iint} \left[(z_{,\eta} - \zeta)^2 + (z_{,\zeta} + \eta)^2 \right] dx\, dy$$

(x, y : cf. Fig. 12. 3-5)· But it is necessary that $z = 0$ is given at each of the axes I to IV .

As a next step we introduce

$$X = \frac{2x}{a}, \quad Y = \frac{y}{x} \quad \text{or} \quad x = \frac{a}{2} X ; \quad y = \frac{a}{2} XY .$$

This yields

$$z_{,\eta} \equiv \frac{\partial z}{\partial \eta} = -\frac{\partial z}{\partial x} = -\frac{\partial z}{\partial X} \cdot \frac{2}{a} + \frac{\partial z}{\partial Y} \cdot \frac{y}{x^2} = \frac{2}{a} \left(-\frac{\partial z}{\partial X} + \frac{Y}{X} \frac{\partial z}{\partial Y} \right),$$

$$z_{,\zeta} \equiv \frac{\partial z}{\partial \zeta} = -\frac{\partial z}{\partial y} = -\frac{\partial z}{\partial Y} \cdot \frac{1}{x} = \frac{2}{a} \left(-\frac{\partial z}{\partial Y} \cdot \frac{1}{X} \right),$$

$$dx = \frac{a}{2}\, dX ; \quad dy = \frac{a}{2} X\, dY .$$

Finally, we introduce $z = \frac{a^2}{4} \cdot \varphi(X, Y)$:

$$z_{,\eta} = \frac{a}{2} \left(-\frac{\partial \varphi}{\partial X} + \frac{Y}{X} \frac{\partial \varphi}{\partial Y} \right); \quad z_{,\zeta} = -\frac{a}{2} \frac{1}{X} \frac{\partial \varphi}{\partial Y} .$$

At last, we state

$$\zeta = \frac{a}{2} - y = \frac{a}{2}(1-XY); \quad \eta = \frac{a}{2} - x = \frac{a}{2}(1-X).$$

So the upper bound theorem reads

$$\hat{I}_t \le \frac{a^4}{2} \overset{①}{\iint} \left\{ \left[-\frac{\partial \varphi}{\partial X} + \frac{Y}{X}\frac{\partial \varphi}{\partial Y} - 1 + XY \right]^2 + \left[-\frac{1}{X}\frac{\partial \varphi}{\partial Y} + 1 - X \right]^2 \right\} X \, dX \, dY. \tag{10}$$

The boundary condition for the true result at line g (Fig. 12.3-5) where $\nu_\eta = 0$, $\nu_\zeta = 1$, reads

$$z_{,\zeta} + \eta = 0 \quad \text{or} \quad \frac{\partial \varphi}{\partial Y} = X(1-X).$$

Thus, the complete set of boundary conditions for $\hat{\varphi}$ in the region ① is :

$$\varphi = 0 \quad \text{if} \quad Y = 1, \; X = 1, \; \text{and} \quad X = 0 \; ;$$
$$\frac{\partial \varphi}{\partial Y} = X(1-X) \quad \text{if} \quad Y = 0 \, .$$

This is satisfied if the function $\varphi = \varphi(X,Y)$ has the form

$$\varphi = X(1-X)(Y-1) + (1-Y^2)(1-X)X \cdot h(Y^2, Y^3, X).$$

As a first approximation we deal with $h \equiv 0$:

$$\varphi = X(1-X)(Y-1).$$

This leads to

$$\hat{I}_t \leq \frac{a^4}{2}\int_0^1\int_0^1 \Big\{\big[(-1+2X)(Y-1)+Y(1-X)-1+XY\big]^2+$$

$$+\big[-1+X+1-X\big]^2\Big\} X\,dX\,dY =$$

$$= \frac{a^4}{2}\int_0^1\int_0^1 \Big\{\big[-Y+1+2XY-2X+Y-XY-1+XY\big]^2+0\Big\} X\,dX\,dY =$$

$$= \frac{a^4}{2}\int_0^1\int_0^1 \Big\{4X^3(Y^2-2Y+1)\Big\}dX\,dY =$$

$$= \frac{a^4}{2}\Big\{4\cdot\frac{1}{4}\cdot\Big(\frac{1}{3}-\frac{2}{2}+1\Big)\Big\} = \frac{a^4}{6},$$

$$\hat{I}_t \leq \frac{a^4}{6} = 0.1667\,a^4.$$

This result is not better than the result for $\varphi \equiv 0$. But if we use $h \equiv \alpha$ as a second approximation we get

$$\hat{I}_t \leq \frac{a^4}{2}\Big\{\frac{1}{3}-\frac{2}{9}\alpha+\frac{4}{15}\alpha^2\Big\}.$$

Minimization of the right side yields

$$\alpha_{opt} = \frac{5}{12}$$

and

$$\hat{I}_t \leq \frac{31}{216}a^4 = 0.1435\,a^4 .$$

Thus, we have calculated by relatively simple means that

$$0.1389\, a^4 \leq \hat{I}_t \leq 0.1435\, a^4 .$$

Looking at any handbook like the German "Hütte I" (28. Aufl., page 926) we see that the true result is $\hat{I}_t = 0.1404\, a^4$.

V) Final remarks

The bounds for $\hat{I}_t$ are integrals. They are quite good. But if we try to calculate $|\tau|_{max}$ from the "best approximation", this value is connected with a derivative and, hence, the error can be relatively large. It is about 10% when we start from the upper bound solution.

13. HYDRODYNAMICS (cf. sect. 4.2 of the lecture)

One problem of viscous flow will illustrate how we can reach upper and lower bounds in hydrodynamics.

Problem 13.-1 : Find out upper and lower bounds for the difference of pressure between two cross-sections 1 and 2 of Fig. 13.-1 in the case of laminar flow of a NEWTON's fluid (η: coefficient of viscosity).

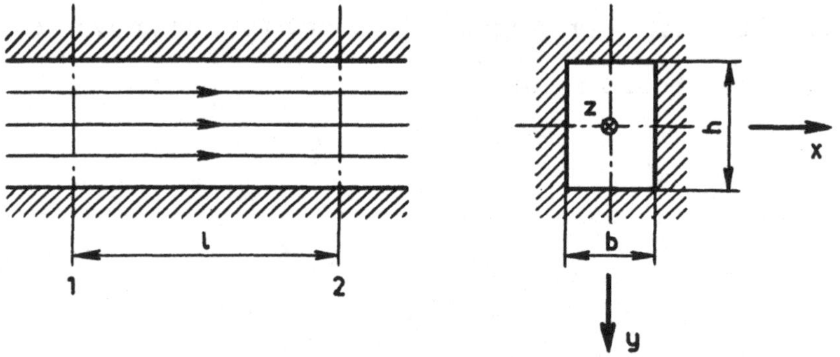

Fig. 13.-1

I) Guessed kinematics and upper bound

We want to use

$$U(\dot{q}) - \Lambda_{ex}(\hat{F} < \dot{x} >) \geq U(\hat{q}) - \Lambda_{ex}(\hat{F} < \hat{x} >),$$

specified as

$$U(\dot{q}) \geq U(\hat{q}) = \frac{1}{2}\Lambda_{ex}(\hat{F} < \hat{x} >) \text{ if}$$

$$\Lambda_{ex}(\hat{F} < \dot{x} >) = \Lambda_{ex}(\hat{F} < \hat{x} >).$$

We cannot state $\dot{x} = \hat{\dot{x}}$. But we can show that $p = -\sigma_{xx} = -\sigma_{yy} = -\sigma_{zz}$ is independent from x and y:

The equations of equilibrium postulate (quasistatical motion):

$$\sigma_{xx,x} + \sigma_{yx,y} + \sigma_{zx,z} = 0\ ; \quad \sigma_{xy,x} + \sigma_{yy,y} + \sigma_{zy,z} = 0\ ;$$

$$\sigma_{xz,x} + \sigma_{yz,y} + \sigma_{zz,z} = 0\ .$$

For a laminar motion, λ_{xy} vanishes, λ_{zx} does not depend on z, λ_{zy} does not depend on z', too. The same is valid for the corresponding shear-stresses $\sigma_{xy}, \sigma_{zx}, \sigma_{zy}$. Hence, the first equations of equilibrium state:

$$\sigma_{xx,x} = 0\ , \quad \sigma_{yy,y} = 0\ .$$

The last equation yields

$$\sigma_{zz,z} = f(x,y)\ .$$

But we know that $\lambda_{xx} = \lambda_{yy} = \lambda_{zz} = 0$, hence, $\sigma_{xx} = \sigma_{yy} = \sigma_{zz}$. These relations are satisfied only if $\sigma_{xx} = \sigma_{yy} = \sigma_{zz} = -p$ is constant in one cross-section. But p is a linear function of z $\quad (\sigma_{zz,z} = -p_{,z} = f(x,y))$.

Thus, we can express Λ_{ex} in the following way (v_i: velocity components) :

$$\Lambda_{ex} = \iint^{A} \Delta p\, v_z\, dA = \Delta p\, \bar{v}\, bh ,$$

where $\bar{v}$ is the average velocity in z - direction. The condition $\Lambda_{ex}(\hat{F} < x >) = \Lambda_{ex}(\hat{F} < \hat{x} >)$ just means $\bar{v} = \hat{\bar{v}}$. This assumption must be made for our upper bound theorem Now, we can state

(1) $$U(\dot{q})\Big|_{\bar{v}=\hat{\bar{v}}} \geq U(\hat{\dot{q}}) = \frac{1}{2}\,\Delta\hat{p}\,\hat{\bar{v}}\, bh .$$

The admissibility condition which the velocity field has to fulfil, is the incompressibility condition $\lambda_{xx}+\lambda_{yy}+\lambda_{zz}=0$. This is given if we estimate

$$v_z = v_z(x,y) \equiv v(x,y) ; \quad v_x = v_y = 0 .$$

The second assumption is that $v_z \equiv v$ has to vanish at two boundaries $x = \pm \frac{b}{2}$, $y = \pm \frac{h}{2}$. Therefore, we estimate $v_z \equiv v$ as

(2) $$v_z \equiv v = v_{max}\left[1-\left(\frac{2x}{b}\right)^2\right]\left[1-\left(\frac{2y}{h}\right)^2\right].$$

The average velocity $\bar{v}$ is then described by

$$\bar{v} = \frac{4}{9}\, v_{max} \quad \text{or} \quad v_{max} = \frac{9}{4}\,\bar{v} .$$

We will have to deal with

$$U(q) = V(Q) = \frac{1}{2}\Lambda_{ex}(F < \dot{x} >) = \frac{1}{2}\Lambda_{in}(Q < \dot{q} >) =$$

$$= \frac{1}{2}\iiint^{V} \sigma_{ij}\lambda_{ij}\,dV = \frac{1}{2}L_{in} \quad (\text{summation of } i, j). \qquad (3)$$

With $\sigma'_{ij} = 2\eta\lambda_{ij}$ (cf the lecture) we get

$$L_{in} = \iiint^{V} 2\eta\,\lambda_{ij}\lambda_{ij}\,dV = \iiint^{V} \frac{1}{2\eta}\,\sigma'_{ij}\,\sigma'_{ij}\,dV. \qquad (4)$$

We can replace λ_{ij} by

$$\lambda_{ij} = \frac{1}{2}\left(\frac{\partial v_i}{\partial x_j} + \frac{\partial v_j}{\partial x_i}\right).$$

The matrix $[\lambda_{ij}]$ of the strain rates λ_{ij} then is

$$[\lambda_{ij}] = \begin{bmatrix} 0 & 0 & \frac{1}{2}\frac{\partial v}{\partial x} \\ 0 & 0 & \frac{1}{2}\frac{\partial v}{\partial y} \\ \frac{1}{2}\frac{\partial v}{\partial x} & \frac{1}{2}\frac{\partial v}{\partial y} & 0 \end{bmatrix}.$$

The inner power L_{in} can be expressed in terms of v by

$$(5) \qquad L_{in}(v) = \eta \iiint^{V} \left[\left(\frac{\partial v}{\partial x}\right)^2 + \left(\frac{\partial v}{\partial y}\right)^2\right] dV .$$

In the estimated velocity field (eq. (2)) we have

$$\frac{\partial v}{\partial x} = v_{max}\left(-\frac{8x}{b^2}\right)\left[1-\left(\frac{2y}{h}\right)^2\right]; \quad \frac{\partial v}{\partial y} = v_{max}\left(-\frac{8y}{h^2}\right)\left[1-\left(\frac{2x}{b}\right)^2\right].$$

Equation (5) then yields

$$L_{in}(v) = \frac{128}{45} v^2_{max} \eta \ell \left(\frac{h}{b} + \frac{b}{h}\right) = \frac{72}{5} \hat{\hat{v}}^2 \eta \ell \left(\frac{h}{b} + \frac{b}{h}\right).$$

The result can be derived from eq (1) :

$$\frac{1}{2} \cdot \frac{72}{5} \hat{\hat{v}}^2 \eta \ell \left(\frac{h}{b} + \frac{b}{h}\right) \geq \frac{1}{2} \Delta \hat{p} \hat{\hat{v}}\, bh$$

or

$$(6) \qquad \frac{\Delta \hat{p}}{\ell} \leq 14.4 \, \eta \ell \hat{\hat{v}} \left(\frac{1}{b^2} + \frac{1}{h^2}\right).$$

II) Guessed statics and lower bound.

We want to apply the theorem

$$(7) \qquad \frac{1}{2} \Lambda_{ex}(\hat{F} < \hat{\hat{x}} >) = U(\dot{q}) \geq \frac{1}{4} \frac{\Lambda_{ex}(F < \hat{\hat{x}} >)}{V(Q)} .$$

Therefore, we need a stress-field to which we can

calculate the external loads as easy as possible. Especially volume forces in z-direction must not appear, because they would act against the unknown velocities v. The easiest way is, to estimate a stress-field with $p = p(z)$ where the equilibrium conditions are satisfied At the boundaries we have $\underline{v} = \underline{0}$;. hence, we do not have to look upon the boundary stresses .

We estimate :

$$\sigma_{xx} = \sigma_{yy} = \sigma_{zz} = -p(z),$$

$$\sigma_{xy} = 0 \; ; \quad \sigma_{xz}, \; \sigma_{yz} \; : \quad \text{function of } x \text{ and } y .$$

$\sigma_{ix,i} = 0$ and $\sigma_{iy,i} = 0$ are satisfied But $\sigma_{iz,i} = 0$ yields with $\tau_i \equiv \sigma_{zi}$:

$$\frac{\Delta p}{\ell} = \frac{\partial \tau_x}{\partial x} + \frac{\partial \tau_y}{\partial y} = \text{const.} \tag{8}$$

We estimate

$\dfrac{\partial \tau_x}{\partial x} = A$; $\dfrac{\partial \tau_y}{\partial y} = B$; hence, $\Delta p = (A+B)\cdot \ell$; $\tau_x = Ax$; $\tau_y = By$.

Equation (7) leads us to :

$$\frac{1}{2} bh \hat{\hat{v}} \Delta \hat{p} = U(\hat{q}) \geq \frac{1}{4} \frac{\left[(A+B)\,\ell\, bh\, \hat{\hat{v}}\right]^2}{\frac{\ell}{2\eta} \iint^{A} (\tau_x^2 + \tau_y^2)\, dA} =$$

$$= \frac{1}{2} \frac{(A+B)^2 \ell\, b^2 h^2 \hat{\hat{v}}^2 \eta}{A^2 \frac{b^3 h}{12} + B^2 \frac{b h^3}{12}}$$

or after having divided by $\frac{1}{2}\, b\, h\, \hat{\bar{v}}\, \ell$:

$$\frac{\Delta \hat{p}}{\ell} \geq 12 \frac{(A+B)^2 \hat{\bar{v}} \eta}{A^2 b^2 + B^2 h^2} = 12\, \bar{v}\, \eta \frac{(1+\alpha)^2}{b^2 + \alpha^2 h^2},$$

where α is the arbitrary quotient $\frac{B}{A}$. Therefore we can optimize (maximize) this lower bound :

$$\frac{\partial}{\partial \alpha}\left(\frac{(1+\alpha)^2}{b^2+\alpha^2 h^2}\right) = 0 \qquad \text{or :}$$

$$2(1+\alpha)(b^2+\alpha^2 h^2) - 2\alpha h^2 (1+\alpha)^2 = 0.$$

This yields $\alpha = b^2/h^2$. Thus, we get the result

$$\frac{\Delta \hat{p}}{\ell} \geq 12\, \hat{\bar{v}}\, \eta \frac{(1+b^2/h^2)^2}{b^2+b^4/h^2} = 12\, \hat{\bar{v}}\, \eta \frac{(h^2+b^2)^2}{h^4 b^2 + b^4 h^2} = 12\, \hat{\bar{v}}\, \eta \frac{(h^2+b^2)^2}{b^2 h^2 (h^2+b^2)}$$

or

$$\text{(9)} \qquad \frac{\Delta \hat{p}}{\ell} \geq 12\, \hat{\bar{v}}\, \eta \left(\frac{1}{b^2} + \frac{1}{h^2}\right).$$

Hence we know

$$12\, \hat{\bar{v}}\, \eta \left(\frac{1}{b^2} + \frac{1}{h^2}\right) \leq \frac{\Delta \hat{p}}{\ell} \leq 14.4\, \hat{\bar{v}}\, \eta \left(\frac{1}{b^2} + \frac{1}{h^2}\right).$$

These bounds are not close together. Therefore, we will try to calculate better upper and lower bounds.

III) Better approximations by an upper bound

In order to get a better upper bound we start from

$$v = v_0\left\{\left[1-\left(\frac{2x}{b}\right)^2\right]\left[1-\left(\frac{2y}{h}\right)^2\right]+\alpha\left[1-\left(\frac{2x}{b}\right)^4\right]\left[1-\left(\frac{2y}{b}\right)^4\right]\right\}.$$

$\bar{v}$ and v_0 are connected by

$$\bar{v} = 4v_0\left\{\frac{1}{9}+\frac{4}{25}\alpha\right\} \quad \text{or} \quad v_0 = \frac{225\,\bar{v}}{4(25+36\alpha)}.$$

By means of eq (5) L_{in} can be calculated to be

$$L_{in} = \frac{128}{25\cdot 7\cdot 9}\, v_0^2\, \ell\eta\left(\frac{b}{h}+\frac{h}{b}\right)\left\{35+96\alpha+80\alpha^2\right\}.$$

Equation (1) yields under these conditions :

$$\Delta\hat{p}\,\hat{\bar{v}}\,bh \leq \frac{128}{25\cdot 7\cdot 9}\;\frac{225^2}{4^2(25+36\alpha)}\,\hat{\bar{v}}^2\ell\eta\left(\frac{b}{h}+\frac{h}{b}\right)\left(35+96\alpha+80\alpha^2\right)$$

or

$$\frac{\Delta\hat{p}}{\ell} \leq \bar{v}\eta\left(\frac{1}{b^2}+\frac{1}{h^2}\right)\cdot\frac{1800}{7}\cdot\frac{35+96\alpha+80\alpha^2}{625+1800\alpha+1296\alpha^2}$$

Optimization of α leads to

$$\alpha_{opt} = 0.221\,,$$

$$\frac{\Delta \hat{p}}{\ell} \leq 14.25\, \eta \bar{v} \quad \frac{1}{b^2} + \frac{1}{h^2}$$

The difference to the first approximation is relatively small.

IV) Better approximation by a lower bound

The lower bound solutions had to start from stress-fields which satisfy

$$\tau_{x,x} + \tau_{y,y} = \text{const.}$$

(eq (8)) This is valid if we assume

$$\tau_x = A\left[x + \beta x^3 - 3\alpha\gamma x y^2\right],$$

$$\tau_y = A\left[\alpha y + \gamma \alpha y^3 - 3\beta x^2 y\right].$$

The calculation based on this assumption is quite complicated because we have to deal with three parameters (α to γ). To simplify the computational work, we restrict this second approximation to $b = h$ so that we can take $\alpha = 1$ and $\beta = \gamma$ because of symmetry.

This yields :

$$\frac{\Delta \hat{p}}{\ell} = 2A \quad \text{and}$$

$$b^2 \hat{\bar{v}}\, \Delta \hat{p}\, \ell \geq \frac{4A^2 \ell^2 b^2 b^2 \hat{\bar{v}}^2 \eta}{\iint\limits^{A} (\tau_x^2 + \tau_y^2)\, dA}$$

with

$$\tau_x = A\left[x + \beta\left(x^3 - 3xy^2\right)\right],$$

$$\tau_y = A\left[y + \beta\left(y^3 - 3x^2 y\right)\right].$$

Introducing $\rho = \beta b^2$, we get

$$\frac{\Delta \hat{p}}{\ell} \geq \frac{\eta \hat{\bar{v}}}{b^2} \cdot \frac{3360}{140 + 28\rho + 9\rho^2}$$

Optimization leads to $\rho_{opt} = \frac{14}{9}$ and

$$\frac{\Delta \hat{p}}{\ell} \geq 28.42\, \frac{\eta \hat{\bar{v}}}{b^2} = 14.21\, \eta \hat{\bar{v}} \left(\frac{1}{b^2} + \frac{1}{h^2}\right) \qquad (h = b).$$

Now we see that it was impossible to reach a better upper bound of the type

$$\frac{\Delta \hat{p}}{\ell} \leq K \cdot \eta \hat{\bar{v}} \left(\frac{1}{b^2} + \frac{1}{h^2}\right),$$

because our results are very close together in the case $b = h$:

$$\underline{\underline{b=h}}: \quad 28.42\,\eta\hat{\hat{v}}\,\frac{1}{b^2} \leq \frac{\Delta\hat{p}}{\ell} \leq 28.50\,\eta\hat{\hat{v}}\,\frac{1}{b^2}.$$

In general we can only say :

$$12\,\eta\hat{\hat{v}}\left(\frac{1}{b^2}+\frac{1}{h^2}\right) \leq \frac{\Delta\hat{p}}{\ell} \leq 14.25\,\eta\hat{\hat{v}}\left(\frac{1}{b^2}+\frac{1}{h^2}\right).$$

The true solution for the other limit case $b \to \infty$ is $\tau_{,y} = A$, $\tau_{,x} = 0$ and yields

$$\frac{\Delta\hat{p}}{\ell} = 12\,\frac{\hat{\hat{v}}\eta}{h^2} = 12\,\hat{\hat{v}}\eta\left(\frac{1}{b^2}+\frac{1}{h^2}\right).$$

If we want to compute better results for general cases we have to start from estimated stress- and velocity-fields with a higher number of free parameters.

14. PLASTOMECHANICS (cf. sect. 4.3 of the lecture)

14.1. Upper and lower bounds to problems of metal forming

In this chapter we will deal with three-dimensional problems of rigid/plastic deformational states. The main yield conditions for this type of processes are connected with the names.

HUBER, LEVY, MISES, HENCKY on one side (mostly called "MISES-criterion") and

TRESCA (TRESCA-criterion) on the other side.

These criteria are in some sense limit cases of the possible criteria if we postulate isotropy and that there is no difference in the behaviour of the material in states of pressure and of tension. Theoretically the MISES-criterion is no bound but practically it is one.

The MISES-criterion states that the yield function f is identical with the second main invariant of the deviators of the stress tensor minus an appropriate function of Y :

$$f \equiv \sum_{i,j} \sigma'_{ij} \sigma'_{ij} - \frac{2}{3} Y^2 = 0 .$$

The corresponding flow rule is

$$\lambda_{ij} = \chi \cdot \sigma'_{ij} .$$

We will not prove that the yield body is convex.

The TRESCA-criterion which we will apply here states that the maximum shear stress is not allowed to exceed $\frac{1}{2}Y$. This can be expressed as

$$f \equiv \sigma_{J\,max} - \sigma_{K\,min} - Y = 0,$$

because τ_{max} is half the difference between the maximum principal stress ($\sigma_{J\,max}$) and the minimum principal stress ($\sigma_{K\,min}$). This can be seen from MOHR's cyrcle. We will restrict our further considerations to this criterion. The flow rule belonging to it postulates first that the principal directions of the stress- and the strain rate tensor coincide because of the isotropy. The second statement is :

$$\lambda_K = 0 \quad \text{if} \quad \sigma_{max} > \sigma_K > \sigma_{min},$$

$\lambda_K > 0$ if $\sigma_{max} = \sigma_K$ and $\lambda_K = -\lambda_J < 0$ if $\sigma_K = \sigma_{min}$, $\sigma_J = \sigma_{max}$ and if the third principal stress differs from σ_K and σ_J.

Difficulties may arise if there are equal principal stresses. This case is very often possible and is called HAAR-v KARMAN criterion despite the fact that it is no criterion but only a special case. The mathematical description is

$$\sigma_J = \sigma_K, \quad \sigma'_L = -2\sigma'_K, \quad J \neq K \neq L \neq J$$

and

$$(\text{sign}\,\lambda_J)\,(\text{sign}\,\sigma'_J) \geq 0\;; \qquad (\text{sign}\,\lambda_K)\,(\text{sign}\,\sigma'_K) \geq 0\;;$$

$$\lambda_L = -(\lambda_J + \lambda_K).$$

This is the mathematical formulation of the field of vectors $\dot{q}$ at a corner of the yield surface.

If we want to apply the upper- and lower-bound theorems of plasticity we need formulae to calculate the power per unit volume and power per unit area at shear surfaces Let us start with the shear zone :

We may have shear (velocity jump) at any surface inside a deformed body. Then, we know that the yield criterion is satisfied with $f = 0$. Therefore, the shear-stress is equal to $\frac{Y}{2}$ and it acts against the maximum of the velocity-jump.

The power per unit volume is given by

$$\sum_{i,j} \sigma_{ij}\,\lambda_{ij} = \sum_{i,j} \sigma'_{ij}\,\lambda_{ij} = \sum_{J} \sigma'_J\,\lambda_J\,.$$

If we have plastic flow and one vanishing principal strain rate, we get :

$$\sum_{J} \sigma'_J\,\lambda_J = \sigma'_{J\,max}\,\lambda_{J\,max} + \sigma'_{J\,min}\,\lambda_{J\,min} =$$

$$= Y\,\text{sign}(\sigma'_{J\,max} - \sigma'_{J\,min}) =$$

$$= Y \cdot \lambda_{J\,max}.$$

In the HAAR-v. KARMAN case we have :

$$\sigma_J = \sigma_K \;, \qquad \sigma_L \neq \sigma_K \qquad (J \neq K \neq L \neq J) \text{ and}$$

$$\sum_J \sigma'_J \lambda_J = \sigma'_J(\lambda_J + \lambda_K) - \sigma'_L(\lambda_J + \lambda_K) =$$

$$= Y \operatorname{sign}(\sigma'_J)(\lambda_J + \lambda_K) =$$

$$= Y\,|\lambda_J + \lambda_K| = Y\,|\lambda|_{max} = w_{in}.$$

Now we are prepared to solve a problem of non-plane motion of a rigid/plastic material.

Problem 14.1-1 : Find out upper and lower bounds for the force P of the axially symmetry extrusion process sketched in Fig 14.1-1 :

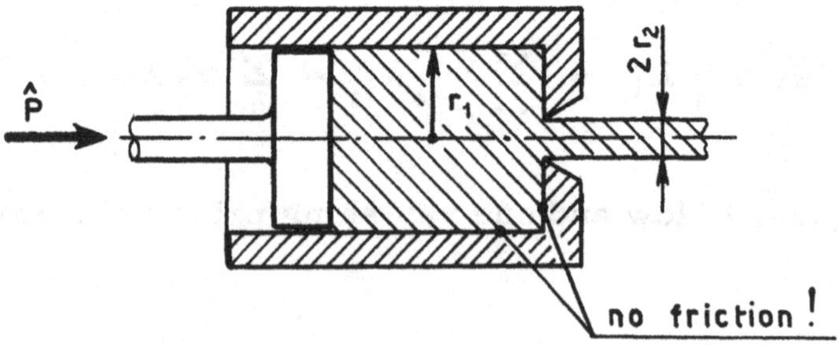

Fig 14.1-1

The assumption that there is no friction is obscure but it is often made

I) Lower bound for $\hat{P}$

To find out lower bounds for $\hat{P}$, we have to estimate an admissible stress field which should be in equilibrium so that it is not necessary to remember volume forces as external effective loads. We separate the zone of deformations into three parts and use the coordinates r and z (cf. Fig. 14.1-2).

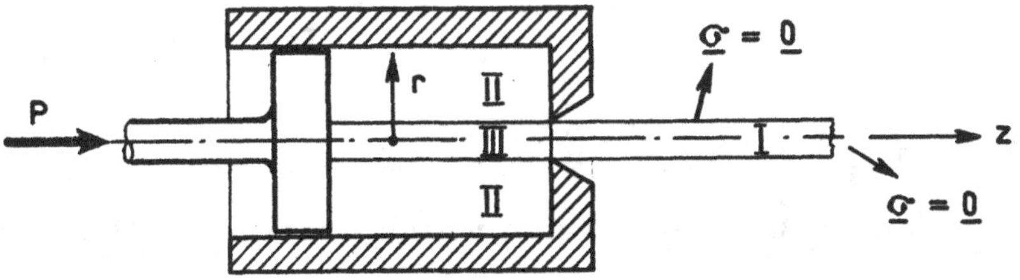

Fig. 14.1-2

If we do not have to deal with volume forces, the only external load is $\hat{P}$. We estimate a field of stresses where the principal axes coincide with the r- and z-directions :

zone I) $\sigma_{rr} \equiv \sigma_r$, $\sigma_{zz} \equiv \sigma_z$, $\sigma_{\vartheta\vartheta} \equiv \sigma_\vartheta$: $\sigma_r = \sigma_\vartheta = \sigma_z = 0$. This is always admissible because there are no external loads at zone I .

zone II) $\sigma_z = 0$ because of the transition between zones I and III , $\sigma_r = \sigma_\vartheta = Y$: the equilibrium condition and the yield condition is satisfied.

zone III) σ_z has to be as low as possible Therefore we state $\sigma_z - \sigma_r = -Y$ and $\sigma_z = \sigma_\vartheta$. Under these conditions, the equations of equilibrium yield

$$\frac{\partial \sigma_r}{\partial r} + \frac{Y}{r} = 0 \; ; \quad \sigma_r = -Y\left(1 + \ln \frac{r}{r_2}\right).$$

This makes possible to have

$$\sigma_z = -Y\left(2 + \ln \frac{r}{r_2}\right)$$

Using $\Lambda_{ex}(\hat{F}\,\hat{y}) \geq \Lambda_{ex}(F\,\hat{y})$ we see :

$\hat{y}$ is the velocity $\hat{v}$ of the punch which prescribes the velocity of the material near the punch Thus, we get

$$\hat{P}\,\hat{v} \geq \hat{v}\int_0^{r_1}(-\sigma_z)\,2\pi r\,dr = \hat{v}\int_{r_2}^{r_1} 2\pi Y\left(2 + \ln\frac{r}{r_2}\right) r\,dr\,,$$

hence,

$$\hat{P} \geq \pi Y \left\{ \frac{3}{2}\left(r_1^2 - r_2^2\right) + r_1^2 \ln \frac{r_1}{r_2} \right\}.$$

II) Upper bound for $\hat{P}$

In order to calculate an upper bound an admissible velocity field must be estimated which should satisfy geometrical boundary conditions. The admissibility condition is the incompressibility. We divide the deformational zone into four parts according to Fig 14.1-3.

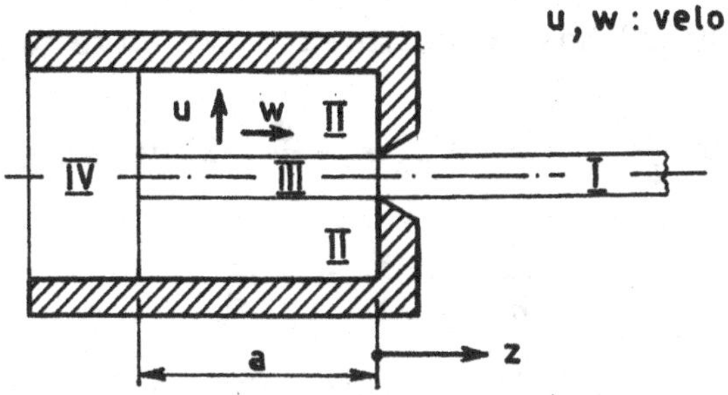

Fig 14.1-3

The estimated velocity-field is :

zone I) $w = \text{const.} = \hat{v}\,\frac{r_1^2}{r_2^2}\,, \quad u = 0$

zone IV) $w = \text{const.} = \hat{v}\,, \quad u = 0$

(zones I and IV: ridig - body - motions)

zone III) $$w = \hat{v}\left[\left(\frac{r_1}{r_2}\right)^2\left(1+\frac{z}{a}\right) - \frac{z}{a}\right] = \hat{v}\left[\left(\frac{r_1}{r_2}\right)^2 + \frac{r_1^2 - r_2^2}{r_2^2}\,\frac{z}{a}\right].$$

These velocities $w(z)$ lead to

$$\lambda_{zz} = \frac{v}{a}\,\frac{r_1^2 - r_2^2}{r_2^2}\,.$$

With $\lambda_{rr} = \frac{\partial u}{\partial r}$, $\lambda_{\vartheta\vartheta} = \frac{u}{r}$, the incompressibility yields $\frac{\partial u}{\partial r} + \frac{u}{r} = -\lambda_{zz}$ with $u\,(r = 0) = 0$.

Thus $u(r)$ is : $u = -\frac{1}{2}\,\hat{v}\,\frac{r}{a}\,\frac{r_1^2 - r_2^2}{r_2^2} = u\,(r)$.

zone II). At the boundary to zone III), u has to be equal to

$$-\frac{1}{2}\hat{v}\,\frac{r_1^2-r_2^2}{a r_2}.$$

If $w = w(z)$ has to be linear, we have to state

$w = -\hat{v}\,\frac{z}{a}$ or $\lambda_{zz} = -\frac{v}{a}$. Thus we get

$$\frac{\partial u}{\partial r}+\frac{u}{r} = -\frac{\hat{v}}{a}.$$

The initial condition to this differential equation is $u = 0$ at $r = r_1$. Thus we get

$$u = -\frac{\hat{v}}{2a}\left(\frac{r_1^2}{r}-r\right).$$

We have to check whether $u(r=r_2)$ has the expected value.

At the boundaries between the different zones, we have some velocity-jumps parallel to the jump-surface. So the admissibility condition is valid.

The next step is to calculate $\Lambda_{in}(\dot{q})$. $\Lambda^{\mathfrak{J}}$ will denote Λ_{in} in zone $\mathfrak{J}$. $\Lambda_{in}^{\mathfrak{J}/k}$ will be Λ_{in} at the boundary between zone $\mathfrak{J}$ and zone K.

$\underline{\underline{\Lambda^{I}}} = 0$

$\underline{\underline{\Lambda^{II}}}$: The principle strain rates are

$$\lambda_{rr} = \frac{\partial u}{\partial r} = \left[\left(\frac{r_1}{r}\right)^2+1\right]\frac{\hat{v}}{2a} > 0,$$

$$\lambda_{\vartheta\vartheta} = \frac{u}{r} = -\left[\left(\frac{r_1}{r}\right)^2 - 1\right] \frac{\hat{v}}{2a} < 0 , \qquad \lambda_{zz} = -\frac{v}{a} < 0 ,$$

hence,

$$|\lambda|_{max} = \lambda_{rr} ,$$

So the result Λ^{II} is :

$$\Lambda^{II} = a \int_{r_2}^{r_1} Y \lambda_{rr} \, 2\pi r \, dr = aY \frac{\hat{v}}{2a} 2\pi \int_{r_2}^{r_1} \left(\frac{r_1^2}{r} + r\right) dr ,$$

$$\Lambda^{II} = \pi \hat{v} Y \left[r_1^2 \ln \frac{r_1}{r_2} + \frac{r_1^2 - r_2^2}{2} \right] .$$

Λ^{III} : The strain rates λ_{rr} and $\lambda_{\vartheta\vartheta}$ are equal, hence : $|\lambda|_{max} = |\lambda_{zz}|$. Thus, we get

$$\Lambda^{III} = \pi \hat{v} Y \left[r_1^2 - r_2^2 \right] .$$

$\Lambda^{IV} = 0$.

So we have reached :

$$\Lambda^{I} + \Lambda^{II} + \Lambda^{III} + \Lambda^{IV} = \pi \hat{v} Y \left[r_1^2 \ln \frac{r_1}{r_2} + \frac{3}{2} \left(r_1^2 - r_2^2\right) \right] .$$

If we would stop here the result would be exactly the lower bound But we have to take care of the jump-surfaces They are : II/IV; III/IV; I/III; II/III.

$\Lambda^{II/IV}$: The velocity jump is given by the amount of u in II :

$$\Delta v = \frac{\hat{v}}{2a}\left(\frac{r_1^2}{r} - r\right).$$

So we get :

$$\Lambda^{II/IV} = \int_{r_2}^{r_1} 2\pi r \, \frac{Y}{2} \, \frac{\hat{v}}{2a} \left(\frac{r_1^2}{r} - r\right) dr =$$

$$= \frac{1}{2} \pi Y \frac{\hat{v}}{a} \left[r_1^2 (r_1 - r_2) - \frac{1}{3}(r_1^3 - r_2^3)\right].$$

$\Lambda^{III/IV}$: Δv is given by u itself inIII The same jump appears at I/III so that we can state $\Lambda^{III/IV} = \Lambda^{I/III}$.

$$\Delta v = |u^{III}| = \frac{1}{2} \hat{v} \frac{r}{a} \frac{r_1^2 - r_2^2}{r_2^2}.$$

This yields

$$\Lambda^{I/III} = \Lambda^{III/IV} = \frac{1}{2} \frac{\hat{v}}{a} \frac{r_1^2 - r_2^2}{r_2^2} \frac{Y}{2} 2\pi \int_0^{r_2} r^2 dr = \frac{1}{6} \pi Y \frac{\hat{v}}{a} \left[r_1^2 - r_2^2\right] r_2.$$

$\Lambda^{II/III}$: Δv is at this jump surface

$$\Delta v = \left(\frac{r_1}{r_2}\right)^2 v \left[1 + \frac{z}{a}\right].$$

Hence we get

$$\Lambda^{II/III} = \int_{-a}^{0} \frac{Y}{2} \Delta v \, 2\pi r_2 \, dz = \frac{1}{2} \pi Y \hat{v} \frac{r_1^2}{r_2} a.$$

The total inner power is

$$\Lambda_{in} = \pi \hat{v} Y \left\{ r_1^2 \ln \frac{r_1}{r_2} + \frac{3}{2}\left(r_1^2 - r_2^2\right) + \right.$$

$$\left. + \frac{1}{2a}\left[r_1^2(r_1 - r_2) - \frac{1}{3}(r_1^3 - r_2^3)\right] + \frac{1}{3a}(r_1^2 - r_2^2)r_2 + \frac{a}{2}\frac{r_1^2}{r_2} \right\} .$$

We can optimize the "parameter" a . In reality we would have to look whether a does not exceed the region on the right side of the punch. We assume that this region is large enough. With the optimal a , we get

$$\hat{P} v \leq \pi v Y r_1^2 \left\{ \frac{3}{2}\left[1 - \left(\frac{r_2}{r_1}\right)^2\right] + \ln \frac{r_1}{r_2} + \sqrt{-\frac{1}{3} + \frac{2r_1}{3r_2} - \frac{1}{3}\left(\frac{r_2}{r_1}\right)^2} \right\},$$

whereas the first result was

$$\hat{P} \geq \pi Y r_1^2 \left\{ \frac{3}{2}\left[1 - \left(\frac{r_2}{r_1}\right)^2\right] + \ln \frac{r_1}{r_2} \right\} .$$

The difference is represented by the square root .

III) Another admissible velocity-field.

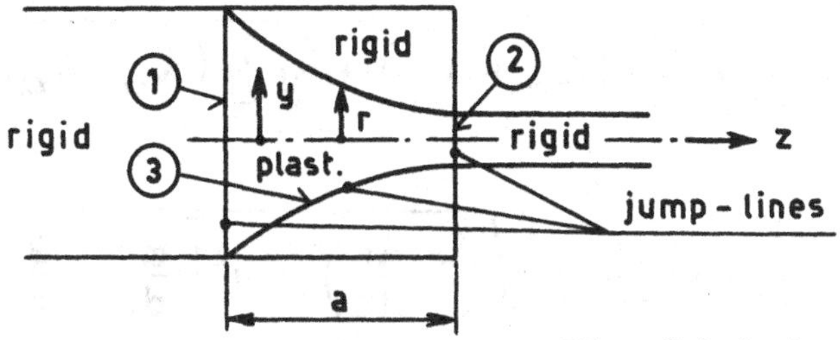

Fig. 14. 1-4

We look at Fig. 14.1-4 :

The plastic zone lies between the jump planes (1) and (2) and is bounded by the third jump line (3) in y-(r-) direction : $r(z)$ describes the surface (3) whereas the general coordinate is y. The length a and the function $r(z)$ are arbitrary and can be optimized.

We postulate that w depends on w only. This is the usual assumption of slab methods. It leads to

$$w(z) = \hat{v}\,\frac{r_1^2}{r^2} \quad \text{or} \quad \lambda_{zz} = \frac{\partial w}{\partial z} = -\frac{2\hat{v}r_1^2}{r^3}\,\frac{dr}{dz}\,.$$

We can apply $\lambda_{rr} = \lambda_{\vartheta\vartheta} = -\frac{1}{2}\lambda_{zz}$ (cf. slab methods)

$$u = \frac{v r_1^2}{r^3}\,\frac{dr}{dz}\,y\,.$$

The calculation of λ_{rz} is more difficult :

$$\lambda_{rz} = \frac{1}{2}\,\frac{\partial u}{\partial z} = \frac{1}{2}\,\hat{v}\,r_1^2\,y\left[\frac{1}{r^3}\,\frac{d^2r}{dz^2} - \frac{3}{r^4}\left(\frac{dr}{dz}\right)^2\right].$$

From MOHR's circle we see

$$|\lambda|_{max} = \frac{1}{2}\left(\lambda_{rr} + \lambda_{zz}\right) + \sqrt{\frac{1}{4}\left(\lambda_{rr} - \lambda_{zz}\right)^2 + \lambda_{rz}^2}$$

$$= \frac{\hat{v}r_1^2}{2r^3}\left\{-\frac{dr}{dz} \overset{+}{\sqrt{\left(3\,\frac{dr}{dz}\right)^2 + y^2\left[\frac{d^2r}{dz^2} - \frac{3}{r}\left(\frac{dr}{dz}\right)^2\right]^2}}\right\}.$$

The boundaries of the ingration are not prescribed if we use z as free coordinate but, if we use r we know the boundaries : r_1 and r_2. Therefore we try to express everything as a function of r. In this sense we handle z as $z(r)$, hence $\frac{dr}{dz} = \left(\frac{dz}{dr}\right)^{-1} = \frac{1}{z'}$.

The second derivative of r can be transformed too:

$$\frac{dr^2}{dz^2} \equiv \frac{d}{dz}\left(\frac{dr}{dz}\right) = \frac{1}{dz}\left(\frac{1}{z'}\right) = \frac{d}{dr}\left(\frac{1}{z'}\right)\cdot\frac{dr}{dz} = \frac{1}{z'}\cdot\left(\frac{1}{z'}\right)' = -\frac{1}{z'^3}z'' = -\frac{z''}{z'^3}.$$

We insert these expressions for $\frac{dr}{dz}$ and $\frac{d^2r}{dz^2}$ into the formula for $|\lambda|_{max}$:

$$|\lambda|_{max} = \frac{\hat{v}r_1^2}{2r^3}\left\{-\frac{1}{z'} + \sqrt{\left(\frac{3}{z'}\right)^2 + y^2\left[\frac{z''}{z'^3} + \frac{3}{r}\frac{1}{z'^2}\right]^2}\right\}$$

Λ^I_{in} will now be calculated as a function of r :

$$\frac{d\Lambda^I_{in}}{dr} = \frac{d\Lambda^I_{in}}{dz}z' = z'\cdot 2\pi Y \int_0^r |\lambda|_{max}\, y\, dy =$$

$$= z'\, 2\pi Y \frac{\hat{v}r_1^2}{2r^3}\int_0^r \left\{yA + y\sqrt{B + Cy^2}\right\}dy.$$

A, B, and C are quantities which do not depend on y, and can be seen from $|\lambda|_{max} = \ldots$ The integration yields :

$$\frac{d\Delta_{in}^{I}}{dr} = z' 2\pi Y \frac{\hat{v} r_1^2}{2r^3} \left\{ \frac{r^2}{2} A + \frac{1}{3C} \cdot \left[\sqrt{B + Cy^2}^{\,3} \right]_0^r \right\} =$$

$$= \pi Y \hat{v} \frac{r_1^2}{r^3} z' \left\{ -\frac{r^2}{2z'} + \frac{1}{3} \left[\frac{z''}{z'^3} + \frac{3}{r} \frac{1}{z'^2} \right]^{-2} \right.$$

$$\left. \times \left[\stackrel{+}{\sqrt{\frac{9}{z'^2} + \left(r \frac{z''}{z'^3} + \frac{3}{z'^2} \right)^2}}^{\,3} + \frac{27}{z'^3} \right] \right\}$$

($z' < 0$ is assumed). This can be rearranged to

$$\frac{d\Lambda_{in}^{I}}{dr} = \pi Y \hat{v} r_1^2 \left\{ -\frac{1}{2r} - \frac{\stackrel{+}{\sqrt{9z'^4 + (rz'' + 3z')^2}}^{\,3} - 27z'^6}{3rz'^2 (rz'' + 3z')^2} \right\} < 0.$$

The integral from r_1 to r_2 is the inner power

$$\Lambda_{in}^{I} = \int_{r_1}^{r_2} \frac{d\Lambda_{in}}{dr} dr = \int_{r_2}^{r_1} -\frac{d\Lambda_{in}}{dr} dr =$$

$$= \pi Y \hat{v} r_1^2 \int_{r_2}^{r_1} \left\{ \frac{1}{2r} + \frac{1}{3r} \frac{\sqrt{9z'^4 + (rz'' + 3z')^2}^{\,3} - 27z'^6}{z'^2 (rz'' + 3z')^2} \right\} dr$$

This is not the total inner power because there are three jump lines ①, ②, and ③. The inner power at ① is :

$$\int_0^{r_1} 2\pi y \frac{Y}{2} |u(y)| dy = \int_0^{r_1} \left(\pi Y y \frac{\hat{v} r_1^2}{r_1^3} \left|\frac{1}{z_1'}\right| y\right) dy =$$

$$= \pi Y \frac{\hat{v}}{r_1} \left|\frac{1}{z_1'}\right| \int_0^{r_1} y \, dy =$$

$$= \frac{1}{3} \pi Y \hat{v} r_1^2 \left|\frac{1}{z_1'}\right| .$$

The power at ③ is similar :

$$\int_0^{r_2} 2\pi y \frac{Y}{2} |u(y)| dy = \int_0^{r_2} \left(\pi Y y \frac{\hat{v} r_1^2}{r_2^3} \left|\frac{1}{z_2'}\right| y\right) dy =$$

$$= \frac{1}{3} \pi Y \hat{v} r_1^2 \left|\frac{1}{z_2'}\right| .$$

At last we have to calculate the power at ② :

$$\Delta u = w \sqrt{1 + \left(\frac{1}{z'}\right)^2} = \hat{v} \frac{r_1^2}{r^2} \sqrt{1 + \left(\frac{1}{z'}\right)^2}$$

The arc length of ② is :

$$ds = dr \sqrt{1 + z'^2} \qquad (dr \overset{!}{>} 0),$$

the area increment dS is then :

$$dS = 2\pi r\, ds = 2\pi r \sqrt{1+z'^2}\, dr .$$

So we compute $\Lambda_{in}^{②}$

$$\Lambda_{in}^{②} = \int^{S} \Delta u \frac{Y}{2}\, dS =$$

$$= \int_{r_2}^{r_1} \left| \hat{v} \frac{r_1^2}{r^2} \sqrt{1+\left(\frac{1}{z'}\right)^2} \cdot \frac{Y}{2} \cdot 2\pi r \sqrt{1+z'^2} \right| dr =$$

$$= \pi Y \hat{v} r_1^2 \int_{r_2}^{r_1} \frac{1}{r} \left| \frac{1+z'^2}{z'} \right| dr =$$

$$= \pi Y \hat{v} r_1^2 \int_{r_2}^{r_1} \frac{1}{r} \left(\left| \frac{1}{z'} \right| + \left| z' \right| \right) dr .$$

Now we use

$$\hat{P} \cdot v \leq \Lambda_{in}(v)$$

and get from $\Lambda_{in} = \Lambda_{in}^{I} + \Lambda_{in}^{①} + \Lambda_{in}^{②} + \Lambda_{in}^{③}$

$$\hat{\varkappa} = \frac{\hat{P}}{\pi r_1^2 Y} \leq \varkappa = \int_{r_2}^{r_1} \frac{1}{r} \left\{ \frac{1}{2} + \frac{1}{3} \frac{\sqrt{9z'^4 + (rz''+3z')^2}^{\,3} - 27z'^6}{(rz''+3z')^2 z'^2} + \right.$$

$$\left. + \left| z' \right| + \left| \frac{1}{z'} \right| \right\} dr + \frac{1}{3} \left(\left| \frac{1}{z_1'} \right| + \left| \frac{1}{z_2'} \right| \right) = F(r, z, z', z'', z_1', z_2') .$$

Optimization of the right side means minimization of F. Thus we get a variational problem :

$$F\left(r, z, z', z'', z_1', z_2'\right) \Longrightarrow \min .$$

The EULER-LAGRANGE equations to this problem are very complicated; therefore we try to find "best approximations" of the optimum by restriction to special curves

case a)

We assume : $z''=0$ or $z'=-d$ $(d>0)$. This simplifies the problem to

$$\varkappa = \ln \frac{r_1}{r_2}\left\{\frac{1}{2}+\frac{\sqrt{1+d^2}-d^3}{d}+d+\frac{1}{d}\right\}+\frac{2}{3d} .$$

The parameter d can be optimized :

$$\left\{\left(2d^2-1\right)\sqrt{1+d^2}-2d^3+d^2-1\right\}\ln\frac{r_1}{r_2}=\frac{2}{3} .$$

It is necessary that $\{\ \} > 0$.

It is easier to calculate $\frac{r_1}{r_2}$ as a function of d than vice versa. We will calculate this function $\varkappa = \varkappa\left(d_{opt}, \frac{r_1}{r_2}\right)$, together with the next case.

case b)

We assume $rz''+3z'=0$ or

$z' = -\frac{c}{r^3} = -b\frac{r_1^3}{r^3}$ (b : optimization parameter).

The relatively large quotient in $F = \ldots$ becomes $\frac{0}{0}$. We have to look for its limit value for $g = rz'' + 3z' \longrightarrow 0$.

$$\lim_{g \to 0} \frac{\sqrt{9z'^4 + g^2}^{\,3} - 27z'^6}{g^2 \cdot z'^2} = \lim_{g \to 0} \frac{\frac{3}{2}2g\sqrt{9z'^4 + g^2}}{2g\,z'^2} =$$

$$= \frac{3}{2} \cdot \frac{3z'^2}{z'^2} = \frac{9}{2}.$$

Hence, we have to state :

$$\varkappa = \int_{r_2}^{r_1} \frac{1}{r}\left\{\frac{1}{2} + \frac{3}{2} + b\frac{r_1^3}{r^3} + \frac{1}{b}\frac{r^3}{r_1^3}\right\}dr + \frac{1}{3b}\left\{1 + \left(\frac{r_2}{r_1}\right)^3\right\} =$$

$$= 2\,\ln\frac{r_1}{r_2} + \frac{b}{3}\left[\left(\frac{r_1}{r_2}\right)^3 - 1\right] + \frac{1}{3b}\left[1 - \left(\frac{r_2}{r_1}\right)^3 + 1 + \left(\frac{r_2}{r_1}\right)^3\right] =$$

$$= 2\,\ln\frac{r_1}{r_2} + \frac{1}{3}\left\{\frac{2}{b} - b\left[1 - \left(\frac{r_1}{r_2}\right)^3\right]\right\}.$$

Optimization yields :

$$\varkappa_{opt} = 2\left\{\ln\frac{r_1}{r_2} + \frac{1}{3}\sqrt{2\left[\left(\frac{r_1}{r_2}\right)^3 - 1\right]}\right\}.$$

The final results are :

$$\hat{\varkappa} \geq \varkappa_1 = \ln \frac{r_1}{r_2} + \frac{3}{2}\left[1 - \left(\frac{r_2}{r_1}\right)^2\right],$$

$$\hat{\varkappa} \leq \varkappa_2 = \ln \frac{r_1}{r_2} + \frac{3}{2}\left[1 - \left(\frac{r_2}{r_1}\right)^2\right] + \sqrt{\frac{2}{3}\frac{r_1}{r_2} - \frac{1}{3}\left[1 + \left(\frac{r_2}{r_1}\right)^2\right]},$$

$$\hat{\varkappa} \leq \varkappa_3 = 2 \ln \frac{r_1}{r_2} + \frac{2}{3}\sqrt{2\left[\left(\frac{r_1}{r_2}\right)^2 - 1\right]},$$

$$\hat{\varkappa} \leq \varkappa_4 = \min\left\{\ln \frac{r_1}{r_2}\left[\frac{1}{2} + \frac{\sqrt{1+d^2}^{\,3} - d^3}{d} + d + \frac{1}{d}\right] + \frac{2}{3d}\right\},$$

where

$$\left[(2 d_{opt}^2 - 1)\sqrt{1 + d_{opt}^2} - 2 d_{opt}^3 + d_{opt}^2 - 1\right] \ln \frac{r_1}{r_2} = \frac{2}{3}$$

Some values of these functions $\varkappa_i$ are given in Table 14.1. -1 (see next page). The results of Table 14.1-1 are printed in Fig. 14.1-5 (see next page).

Table 14.1-1

$\frac{r_2}{r_1}$	$\frac{r_1}{r_2}$	$\varkappa_1$	$\varkappa_2$	$\varkappa_3$	$\varkappa_4$	d_{opt} (≈)
0.9999	1.0001	0.0004	0.0120	0.0155	0.0165	81.7
0.999	1.001	0.0040	0.0405	0.0536	0.0540	25.8
0.990	1.01	0.0395	0.1546	0.1839	0.1842	8.2
0.952	1.05	0.1886	0.4425	0.4718	0.4728	3.83
0.909	1.10	0.3557	0.7083	0.7330	0.7145	2.88
0.833	1.2	0.6407	1.126	1.169	1.165	2.21
0.667	1.5	1.239	1.966	2.265	2.180	1.74
0.5	2	1.818	2.776	4.020	3.439	1.55
0.25	4	2.792	4.313	9.289	6.427	1.40
0.10	10	3.788	6.308	34.408	10.355	1.33

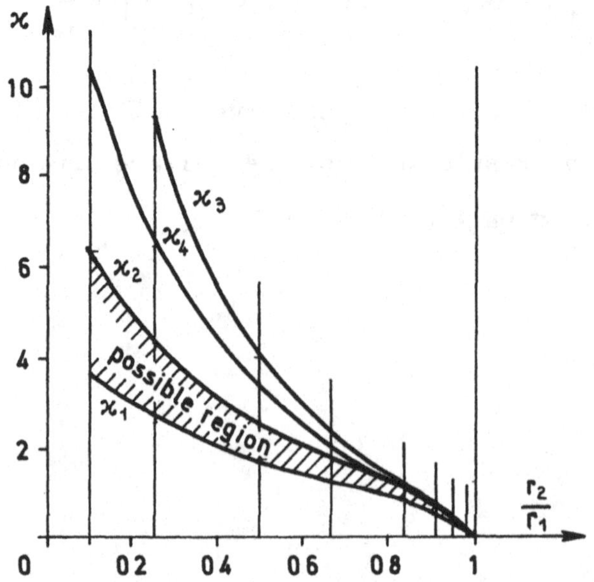

Fig. 14.1-5

14.2. Load carrying capacity

Civil engineers who want to calculate the load which can be carried by any system often use the method of plastic limit-design. The idea for a system consisting of bars only is pointed out in the lecture course. Beams can be treated in a similar way.

We assume that the beam is either rigid or plastic If, at any point, the bending moment reaches a value M_0 where M_0 is given by the fact that $\sigma_b = \pm Y$ and $\int^A \sigma_b \, dA = 0$, the beam behaves as if it would contain a hinge with a moment of dry friction M_0 at this point. M_0 depends on Y and on the geometry of the cross-section. In reality, it depends on the normal-force, too but we will neglect this influence.

Under these circumstances admissibility of a force-field means : the bending-moment does not exceed the value M_0. Such an admissible force field can be used for the calculation of a lower bound of the load carrying capacity. We will illustrate this by an example.

Problem 14.2-1 : Find out a lower bound and - afterwards - the true value of the load carrying capacity for the frame sketched in Fig. 14.2-1. P be the only external load.

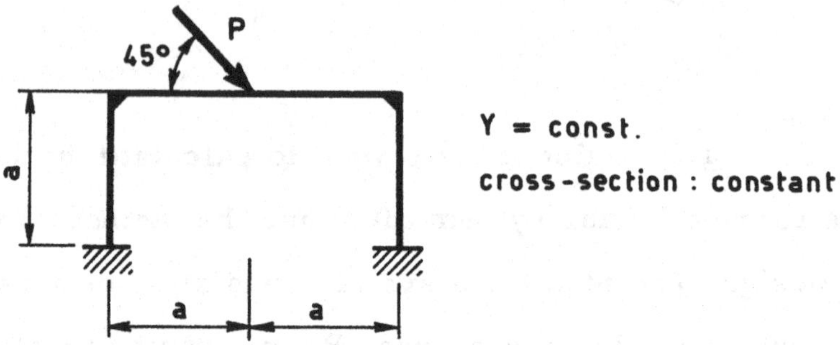

Fig. 14. 2-1

$\hat{P}$ will denote the true value of the load carrying capacity.

An admissible statical field is given in Fig. 14. 2-2. It belongs to a system with two hinges at the points A and B .

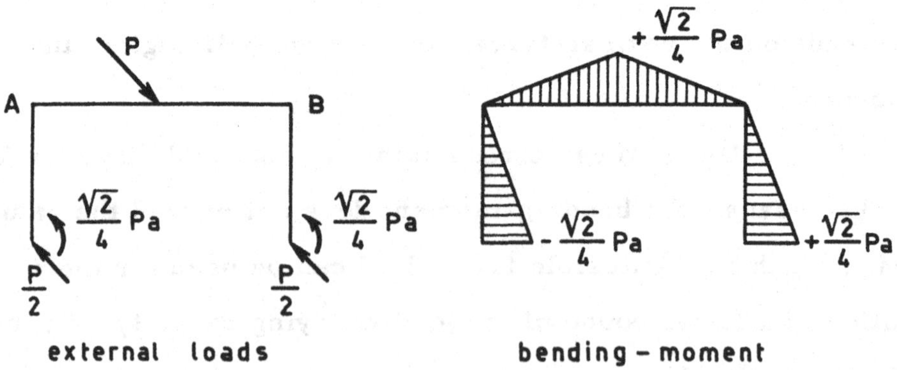

Fig. 14 2-2

If the maximal bending-moment is taken to be equal to M_0, we get $\frac{\sqrt{2}}{4}\,Pa = M_0$ or $P = 2\sqrt{2}\,\frac{M_0}{a}$.

On the other hand,

$$\Lambda_{ex}\left(\hat{F} < \hat{\underline{y}} >\right) \geq \Lambda_{ex}\left(F < \hat{\underline{y}} >\right)$$

can simply be applied because of $\hat{\underline{y}}_P \equiv \hat{\underline{y}}_{\hat{P}}$. Hence, we can formulate

$$\hat{P}\hat{\underline{y}} \geq P\hat{\underline{y}} \quad \text{or} \quad \underline{\underline{\hat{P} \geq P = 2\sqrt{2}\,\frac{M_0}{a}}}\,.$$

So we have calculated a lower bound for the load carrying capacity.

If we want to determine the true value of the load carrying capacity we have to make the system statically determinate by introducing three reaction forces as if they were external loads. We can easily imagine that plastic hinges can only appear at points where M_b is able to be a maximum or a minimum. But M_b is linear. Therefore these points are corners of the frame and points where external loads are acting. Fig. 14.2-3 shows the five possible plastic hinges and the external loads which will be used later on.

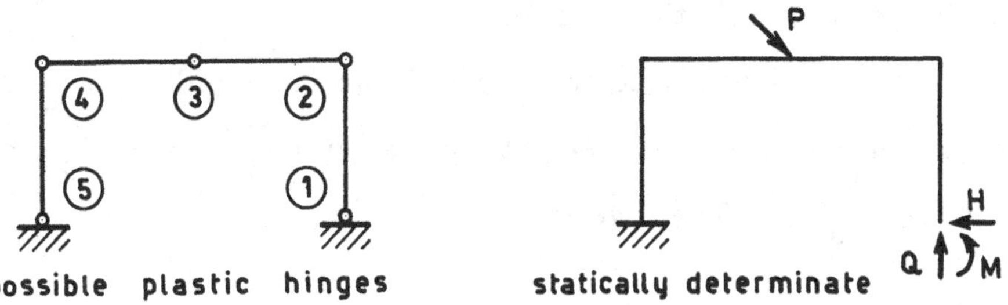

possible plastic hinges

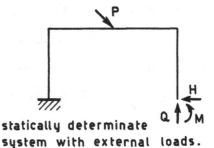

statically determinate system with external loads.

Fig. 14.2-3

The bending-moment at the points ① to ⑤ can easily be calculated. We have to state that the amount of M_b must not exceed M_0. So we derive the following inequalities :

① : $|M| \leq M_0$

② : $|M - Ha| \leq M_0$

③ : $|M - Ha + Qa| \leq M_0$

④ : $|M - Ha + 2Qa - (\sqrt{2}/2)Pa| \leq M_0$

⑤ : $|M + 2Qa - \sqrt{2}\,Pa| \leq M_0$.

Each of these inequalities represents two linear boundary-relations for the admissible values of M, H, Q, and P. We search for an optimal result where the optimum means : a maximal P. It depends linearly on M, H, Q, and P, too. Hence we can apply the methods of linear programming which were presented in section 1.2.

We have to deal with 10 conditions concerning four quantities. Hence, 210 combinations are possible. So we are forced to use the iteration method.

Before doing so we write these ten conditions in a more convenient manner. We introduce $m = \frac{M}{a}$, $p = \frac{\sqrt{2}}{2} P$, and $m_0 = \frac{M_0}{a}$. Then we get :

1 a) $m \le m_0$

1 b) $-m \le m_0$

2 a) $m - H \le m_0$

2 b) $-m + H \le m_0$

3 a) $m - H + Q \le m_0$

3 b) $-m + H - Q \le m_0$

4 a) $m - H + 2Q - p \le m_0$

4 b) $-m + H - 2Q + p \le m_0$

5 a) $m + 2Q - 2p \le m_0$

5 b) $-m - 2Q + 2p \le m_0$

We want to reach the result with a small number of steps. Therefore, we estimate the possible motion of the system.

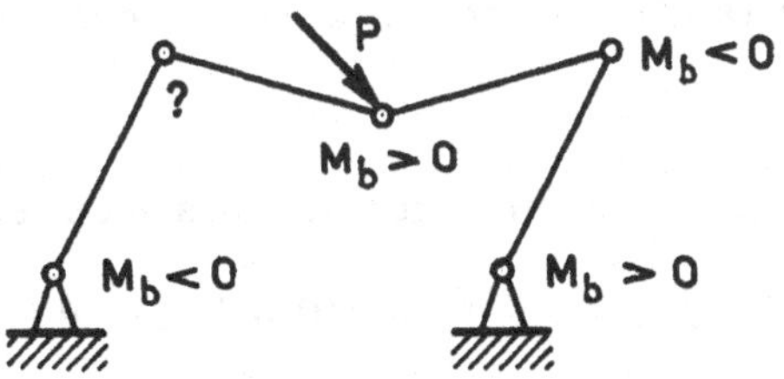

Fig 14.2-4

Fig. 14.2-4 presents the estimated mode of motion. We see that conditions 1a, 2b, 3c, 5c, and, possibly, 4a, 4b will be of interest. But we do not know the sign of the bending-moment at point (4). Hence, this point seems to be not so interesting.

So we start from the combination 1a, 2b, 3c, 5b :

The equalities (cf. sect. 1.2) are :

1a) $m = m_0$,

2b) $H - m = m_0$,

3a) $m - H + Q = m_0$,

5b) $-m - 2Q + 2p = m_0$.

They yield : $m = m_0$; $H = 2m_0$; $Q = 2m_0$; $p = 3m_0$,

and :

$$P = \sqrt{2}\,p = 3\sqrt{2}\,\frac{M_0}{a}$$

Now we have to check whether two conditions are satisfied :

1. The calculated point has to be a corner of the admissible region.

   This is given if the other conditions are satisfied (with an inequality sign), too. We remember that two corresponding conditions (for example 1a and 1b) are satisfied if one of them is fulfilled with an equality sign. Hence, in our case, conditions 1b , 2a , 3b and 5a are satisfied. We have to check conditions 4a , 4b :

$$|m - H + 2Q - p| = |m_0 - 2m_0 + 4m_0 - 3m_0| = 0 \le m_0 .$$

We see :

<u>The calculated point is a corner of the admissible region</u>

2. P has to be a lower value than $3\sqrt{2}\,\frac{M_0}{a}$ at each of the "neighbouring-points" (cf. sect. 1.2).

The combination 1a, 2b, 5b leads to a neighbouring point satisfying the equations 1a, 2b, 5b, 4b :

$$m = m_0 \; ; \quad Q = m_0 \; ; \quad H = 2m_0 \; ; \quad p = 2m_0 < 3m_0 \, .$$

The other directions are counted like a table :

$$1a, 2a, 5b \quad + \quad 2a \; : \; m = m_0 \; ; \; H = Q = 0 \; ; \; p = m_0 < 3m_0$$

$$2b, 3a, 5b \quad + 4a \; : \; m = -m_0 \; ; \; Q = 2m_0 \; ; \; H = 0 \; ; \; p = 2m_0 < 3m_0 ,$$

$$1a, 2b, 3a \quad + \quad 4a \; : \; m = m_0 \; ; \; Q = H = 2m_0 \; ; \; p = 2m_0 < 3m_0$$

We recognize : $m = m_0$, $H = Q = 2m_0$, $p = 3m_0$ represents the optimum. Hence,

$$\hat{P} = 3\sqrt{2}\,\frac{M_0}{a} \, .$$

The statics belonging to this result are printed in Fig 14.2-5

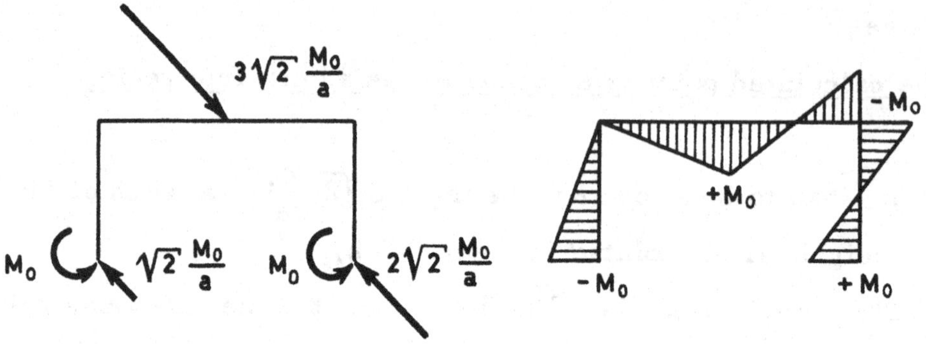

Fig. 14.2-5

At last, one question remains. It cannot be solved without taking into account the elastic deformations :

When a system is loaded by a growing load P, does really the system reach the optimum state of internal loads? Special considerations show that, indeed, the optimum will always be reached, but they cannot be point out here.

Table of figures, problems and tables

Figures

Problems

Tables

CONTENTS

Printed in the United States
By Bookmasters